Agriculture - Environment - Medicine

Edited by
Katsu Minami

Kitasato University Agromedicine Symposia
2006-2008 Abstracts

ISBN 978-4-8425-0454-4

Cover :
 Photographer   Hiroshi Isogai " NIPPON SEEN FROM THE AIR " Yama-Kei Publishers Co., Ltd.
 Courtesy   Shimizukobundo Publishing Co.

Copyright 2009 Kitasato University Office of the President
All rights reserved
Printed in Japan by YOKENDO Publishers, Tokyo

Contents

Prerface ·· iii

Part I   Agriculture, Environment and Healthcare ································· 1

Part II   Alternative Medicine and Alternative Agriculture ······················· 37

Part III A Look at Avian Influenza from the Perspective of Agriculture,
Environment, and Medicine ··· 69

Part IV Effect of Cadmium and Arsenic on Agriculture, the Environment, and Health ·· 103

Part V   Global Warming: Assesing the Impacts on Agriculture, the Environment,
and Human Health, and Techniques for Responding and Adapting ·········· 137

Part VI Food Safety and Preventive Medicine ·································· 167

Contributors ··· 221

Agriculture – Environment – Medicine

Preface

The goals of preventive medicine in the 21st century include the management, assessment, and communication of risk, prevention of disease, and improvement of health. To address the expectations of society today, it is vital to investigate and establish common ground between these medical issues and agriculture.

The outcomes of 20th century scientific and technological advances suggest very strongly that research, education, and extension in agromedicine will be absolutely vital to human society in the 21st century to prevent disease, promote health, ensure food safety, practice environmentally friendly agriculture, benefit from the therapeutic value of agriculture, and otherwise ensure human happiness. If the statement, "We are what we eat," is true, then I feel we have not paid enough attention to agromedical research, education, and extension.

A major problem in modern society is disjunction in various forms – disjunction between individuals, teachers and students, the soil and people, facts, culture and history, and the present and the past. These disjunctions can be roughly divided into four categories: disjunction between knowledge from knowledge, between knowledge and action, between knowledge and feelings, and between past and present knowledge. Agriculture and medicine are also disconnected, and surmounting this disconnection will require an all – embracing, multidisciplinary approach to research and education in areas of overlap between agriculture and medicine.

With this goal in mind, Kitasato University held symposiums on agromedicine and this book summarizes the abstracts of the following symposiums: "Agriculture, environment and healthcare," "Alternative medicine and alternative agriculture," "A look at avian influenza from the perspective of agriculture, environment, and medicine," "Effect of cadmium and arsenic on agriculture, the environment, and health," "Global Warming: Assessing the impacts on agriculture, the environment, and human health, and techniques for responding and adapting," and "Safety of food and preventive medicine."

I wish to dedicate this book to the primary care providers, agronomists, environmentalists, and health professionals who seek more knowledge of the agricultural, environmental, and medical problems they manage. I hope this book fulfills your needs and that it becomes a useful tool for translating complex agricultural information into sound advice for medicine, and conversely, for translating complex medical information into sound advice for agriculture.

I would like to thank Mr. Tetsuya Furuya and Miss Etsuko Tanaka. Their organization and editorial assistance were important in putting this book together.

Katsu Minami
Kitasato University

Part I
Agriculture, Environment and Healthcare

1   A Message from the Symposium Organizer ··················· 3
      Tadayoshi Shiba

2   The Need for Collaboration between Agriculture, Environment and Healthcare
      ·· 5
      Katsu Minami

3   Chiba University Center for Environment, Health and Field Studies :
      Philosophy and Practice ·· 11
      Toyoki Kozai

4   Agromedicine from the Perspective of Medicine ···················· 17
      Yoshiharu Aizawa

5   Food, Agriculture, and the Environment ·························· 21
      Isoya Shinji

6   Combining Oriental Medicine with Horticultural Therapy ················ 27
      Toshiaki Kita

7   Human Health and Functional Foods ····························· 31
      Takafumi Kasumi

Chapter 1

A Message from the Symposium Organizer

Tadayoshi Shiba

Allow me to make a few remarks as representative of the organizer in relation to Kitasato University's 1st Agromedicine Symposium.

One of the trends of modern science is a move away from specialization (partial optimization) towards a more holistic approach (overall optimization). In the past, this was not always the case. For example, the increasing segmentation of a university into individual faculties and departments was considered vital to the development of various fields of study. In physics for example, this approach led to the identification of quarks as the ultimate building blocks of matter, while in biology this approach led to dramatic advances being made in elucidation of cells at the DNA level. This reductionism — breaking things down into ever smaller, ever more basic units — served as the backbone of modern science and, particularly in the 20th century, produced astounding results. However, from the second half of the 20th century, we have witnessed the emergence of an array of issues of global scale — environmental pollution, resource depletion, famine, population explosion, and so forth — that reductionism is ill — equipped to address. While these each have their own fields, they are all part of the same chaotic current; looking at each issue individually would be like investigating individual waves and eddies in the flow of a huge river, and failing to grasp the underlying forces involved. In other words, reductionism with its endless segmentation and compartmentalization can cause one to lose sight of the ties between different issues and at times fall into the trap of drawing dichotomies and seeing things in terms of binary oppositions. The vicious circle that arose in 20th century society between economic growth and environmental degradation is a prime example. What is required is an ecological approach that looks at relationships between the different components of a single ecosystem. And in the present time of increasing interplay between science and technology, the integration of science with ethics is also a crucial requirement. Environmental ethics and bioethics are 2 key examples. I feel that it is only natural for science in the 21st century not so much to reject reductionism, but rather to seek to go beyond it to arrive at a new intellectual paradigm.

Such is my basic understanding of the current shift in intellectual paradigms. At this university, in the tradition set by our founder Shibasaburo Kitasato, we are beginning to integrate agriculture, environment, and medicine as a model for the development of a unified intellectual approach. It is my fervent hope that this symposium will serve as a forum for valuable and pragmatic discussion that gives birth to new ideas and hints for a higher-level holistic approach to food, environment, and health issues. I would also like to express my heartfelt thanks to Chiba University and Tokyo University of Agriculture for their support of this event, and to all of the Nihon University faculty who have helped.

Chapter 2

The Need for Collaboration between Agriculture, Environment, and Healthcare

Katsu Minami

Foreword

Geared to mass production and economic efficiency, modern agriculture has become an intensive system that, with the use of chemical fertilizers, agricultural chemicals, and synthetic materials, has been able to produce large quantities of food for the growing human population. However, the enormous resources and energy poured into this system of mass production have caused environmental headaches of various kinds, from localized problems such as heavy metal contamination, to more widespread problems such as the eutrophication of rivers, lakes, and marshes due to runoff of nitrogen and phosphorus, and global warming caused by methane and N_2O emissions. More recently, problems such as dioxin emissions with impacts that span generations have emerged, bringing grave implications for human health and the global environment.

Along with progress in fields of medicine such as microbiology, immunology, clinical medicine, and pharmacology, advances in nutrition science have enabled many people to surmount illness and improve their health. At the same time, the many different chemical substances invented or discovered in the process have given rise to drug-related diseases and other problems, providing insights that have driven the further evolution of clinical medicine. There are also unresolved issues related to "human healing" and so forth.

The goals of preventive medicine in the 21st century include the assessment, management, and communication of risk, prevention of disease, and the improvement of health. To address the expectations of society today, it is vital now to investigate and establish common ground between these medical issues and agriculture.

The outcomes of 20th century scientific and technological advances suggest very strongly that research and education in agromedicine will be absolutely vital to human society in the 21st century. The need for such research and education to prevent disease, promote health, ensure food safety, practice environment-friendly agriculture, benefit from the therapeutic value of agriculture, and otherwise ensure human happiness cannot be overstated. Considering the veracity of the saying "We are what we eat", I feel that not enough attention has been paid to agromedical research and education.

A major problem in modern society is disjunction in its various forms-disjunction between people, between teachers and students, between soil and nature and people, between facts, between culture and history and the present time, and so forth. The list could go on and on, but disjunction illness could be divided roughly into 4 categories: disjunction between knowledge from knowledge, between knowledge and action, between knowledge and feelings, and between past and present knowledge. Disjunction exists also between agriculture and medicine, and surmounting this disjunction requires an all-embracing, multi-disciplinary approach to research and education in areas of overlap between agriculture and medicine.

World trends

Kitasato University has long dedicated itself to the comprehensive pursuit of life sciences, but agromedicine is nevertheless a relatively new field for us. Overseas, the North American Agromedicine Consortium (NAAC) has been publishing the Journal of Agromedicine (http://www.haworthpress.com/web/JA/) and a newsletter since it was established in 1988, and there are a number of other publications which I will skip owing to lack of space. I want to look here at a number of different international initiatives and topics in agromedicine, and try to identify the underlying trends in this discipline.

1) International Nitrogen Initiative (INI)

A century has gone by since the Haber-Bosch process was first used to fix atmospheric nitrogen, and currently 270 Tg (10^{12} g) of nitrogen is being fixed every year. Surplus fixed nitrogen is beginning to have major impacts on the natural environment and human communities, such as nitrate-polluted groundwater, eutrophication, destruction of the ozone layer, global warming, blue baby syndrome, nitrate-polluted crops and livestock, effects on human health, and so forth. There is an urgent need for research on the effects on human health of surplus nitrogen in food and the environment, and for the implementation of appropriate countermeasures. The International Nitrogen Initiative was launched to address this issue. (http://www.initrogen.org/)

2) European Nutrigenomics Organization (NuGO)

Nutrigenomics is the study of how nutrients from food affect the expression and regulation of genes. Nihon University's Professor Takafumi Kasumi gives a detailed presentation of the field in his talk. There has never been a greater need than there is now for research in combined fields, and nutrigenomics is one such combined field. Nutrigenomics combines elements of genomics, computer science, immunology, pathology, agriculture, oceanography, environmental science, analytical chemistry, and life sciences in relation to various foodstuffs to assess health, illness, and stages in between from the viewpoint of diet. As such, nutrigenomics is a combined field that feeds into agromedicine.
(http://www.ifr.bbsrc.ac.uk/Science/ScienceBriefs/nugo.html)

3) Persistent organic pollutants (POPs)

POPs are not only highly toxic, but also degrade only very slowly in the environment, and tend to accumulate within living organisms. POPs circulate through the atmosphere, soil, water, and organisms to contaminate manufactured foodstuffs. Currently aldrin and eleven other substances are subject to the Stockholm Convention on POPs. (http://www.pops.int/)

4) Codex Alimentarius Commission

Created jointly by the FAO and WHO, the Codex Alimentarius Commission is an international food standards organization dedicated to protecting the health of consumers and ensuring fair trade practices in the food trade. For example, the Commission's Ad Hoc Intergovernmental Task Force on Food Derived from Biotechnology (host country: Japan) handles such issues as latent risks to human health of consuming food products derived from genetically modified (GM) animals, food products derived from GM plants that can contribute to nutrition or health, GM adulteration, and GM plant-derived foods containing pharmacological ingredients and bioactive substances.
(http://www.codexalimentarius.net0/,http://www.n-shokuei.jp/codex/)

5) Global Environmental Change and Human Health (GECAHH)

One of the components of the International Geosphere − Biosphere Programme (IGBP) is the Earth System Science Partnership (ESSP). ESSP activities include joint projects, one of which, Global Environmental Change and Human Health (GECAHH), emphasizes the need for linkage between agriculture, environment, and medicine.
(http://www.igbp.kva.se/cgi − bin/php/frameset.php)

6) Avian influenza virus (AIV)

AIV was up until recently carried naturally by waterfowl, curlews, and plovers, but owing to factors such as international trade and the industrialization of poultry farming, AIV ecosystems, distribution, range of hosts, and pathogenicity have changed extensively. The global pet bird trade; duck farms; open-range poultry farms; the transport and sale of live poultry, pet birds, and fighting cocks; and the increasing scale of poultry farms have all played a part. According to the WHO, as of July 20, 2006, 230 persons in 10 countries had been infected by H5N1 AIV, with 133 − almost half − dying as a result.
(*Joho: Nou − Kankyo − Iryo* [Newsletter: Agriculture, Environment, and Medicine] No.-16, pp. 10-11, pub. Kitasato University)

Domestic (Japanese) trends

1) Chiba University Center for Environment, Health, and Field Studies

Since Chiba University President Toyoki Kozai introduces the Center's activities in his talk, I will skip details here. (*Joho:Nou − Kankyo − Iryo* [Newsletter: Agriculture, Environment, and Medicine] No.-1 pp. 10-12, No. 2 pp. 4-5, pub. Kitasato University)

2) Shimane University Medicine-Industry-Agriculture Project for a Long and Healthy Life

Committed to research in areas of overlap between medicine, industry, and agriculture Shimane University has launched its "Medicine-Industry-Agriculture Project for a Long and Healthy Life-Developing a New System for Analyzing Human Health and Functional Foods Derived from Local Produce", a journal for which is already being published.
(*Joho: Nou − Kankyo − Iryo* [Newsletter: Agriculture, Environment, and Medicine] No. 8

pp. 18-19, pub. Kitasato University)

3) Science Council of Japan: Consolidation of 7 divisions into 3

The Science Council of Japan's 3 humanities divisions (Literature, Law Economics) and 4 natural sciences divisions (Physical Sciences, Engineering, Agriculture, Medicine) have been consolidated into 3 divisions (Humanities, Life Sciences, Physical Sciences and Engineering), and jurisdiction has been transferred from the Minister of Public Management, Home Affairs, Posts and Telecommunications to the Prime Minister. The consolidation of physical sciences, agriculture, and medicine into a new Life Sciences division signals the need for collaboration between agriculture, environment, and medicine.

(*Gakujutsu no Doko* [Trends in Academia] November 2005;*Joho : Nou — Kankyo — Iryo* [Newsletter: Agriculture, Environment, and Medicine] No. 9 pp. 3-4, pub. Kitasato University; Science Council of Japan website: http://www.scj.go.jp/en/)

4) Animal-assisted therapy (AAT) and animal-assisted activity (AAA), companion animals

Topics include the influence of companion animals on child development, companion animals and elderly people, influence of companion animals on human physiology, and the use of companion animals in therapy.

(*Joho : Nou — Kankyo — Iryo* [Newsletter: Agriculture, Environment, and Medicine] No. 5 pp. 8-15, pub. Kitasato University)

5) 2005 White Paper on Agriculture: Food Safety, Health, and Medicine

The Ministry of Agriculture, Forestry and Fisheries submits a white paper to the Prime Minister every year, and its 2005 White Paper on Food, Agriculture, and Rural Areas contains a number of new items pointing to the need for collaboration between food safety (agriculture) and health (medicine), including: "Ensuring Food Safety and Consumer Confidence: BSE and Highly Pathogenic Avian Influenza", "Food in Relation to Health and Healthcare", "Improving Dietary Habits", "Dietary Education", "Initiatives to Ensure Food Safety", and "New Initiatives for Rural Areas Focused on Health and Welfare".

Affinity between agriculture and medicine

Agriculture and medicine were once intimately tied, and are taking the same path even today. The following is an outline of this affinity from a historical perspective.

(1) It can be surmised from burial remains when people started to use rituals, and it was from these rituals that possibilities for the medical act of cleansing emerged. Agriculture, too, demanded the use of rituals to beseech the gods for the protection of crops from storms, drought, and other calamities.

(2) As civilization emerged, it gave rise to the study of medicine. Civilization in turn owes its emergence to agriculture, and in what was a synergic relationship, civilization prompted the further growth of agriculture as it developed.

(3) Hippocrates established medicine as a discipline that has served humankind ever since by teaching that disease is not caused by supernatural forces, but is rather a natural phenomenon that can be understood by utilizing rational concepts and learning from experience. In

the same way, people came to learn that wheat and barley, which are among the most ancient grain crops, could be cultivated practically, rather than just left to propagate naturally.

(4) Religions such as Confucianism, Taoism, Buddhism, Hinduism, Christianity, and Islam emerged to have immeasurable impacts on medicine, both materialistically and spiritually. Various types of agriculture, including Mediterranean, savannah, root cultivation, New World, and rice cultivation, also emerged to influence the science of agriculture.

(5) The "live" fields of surgery, anatomy, and physiology developed during the European Renaissance, leading to a flowering of hospital-based medical care. Meanwhile, field studies led to the development in European agriculture of the 3-field system and other crop rotation systems.

(6) The quality of life and health of workers deteriorated under the capitalism of the Industrial Revolution, prompting the rapid development of public hygiene and social medicine studies. Crop rotation enabled the feeding of the Industrial Revolution's growing urban population. Landowners adopted the Norfolk system of crop rotation in increasing numbers.

(7) Lab-based medical research focused on the cause and prevention of epidemics began to develop in earnest from the latter half of the 19th century. In 20th century, Biochemistry merged with molecular biology to become a powerful tool for investigating life processes. In agriculture, crop yields increased considerably as the manufacture and use of chemical fertilizers and agricultural chemicals took off. Later advances in molecular biology led to the birth of genetically modified crops.

(8) Just as there is alternative agriculture, there is also alternative medicine. The former covers alternatives to agrochemical-based intensive cultivation methods, while the latter covers alternative treatment methods to the "Western" medicine that has become the mainstream today. Both alternative agriculture and alternative medicine incorporate the holistic approach of life sciences.

(9) The mapping of both the human genome (medicine) and the rice genome (agriculture) was completed at the start of the 21st century.

Cooperation between agriculture and medicine

As explained above, agriculture and medicine share the same roots and have followed similar paths historically. They also both have a vital role to play in today's disjointed society. The following is a list — in no particular order — of areas of overlap and cooperation between agriculture and medicine:

Physiology, endocrinology, nutrition and vitamins, infectious diseases, biochemistry and molecular biology, environmental pollution, medicinal plants, use of eco-friendly agricultural produce, functional foods, animal therapy, diagnosis with positron emission tomography (PET), and facilities such as Kitasato Research Center of Environmental Sciences that will become increasingly essential for the way they enable integrated agricultural, environmental, and medical analysis.

Closing comments

The 11th chapter of the Tao Te Ching, a seminal ancient Chinese work said to have been penned by philosopher Lao Tsu, the father of Taoism, contains the following words:

"Thirty spokes join together in a wheel, but it is the centre hole that makes the wagon move. We shape clay into a pot, yet it is the emptiness that we use. We fashion wood for a house, but it is the inner emptiness where we live, and the empty holes of windows and doors that makes it livable. Therefore, being is what we have, but non‐being is what we use."

This passage points to a fundamental principle for integrating diversity. Put another way, it could be interpreted as expressing the essence of the relationship between agriculture, environment, and medicine; or between food, soil and water, and health. The lump of clay, or the windows and doors represent specialization and differentiation, whereas the wheel, pot, and house represent the integration of diversity. You could say that the presentations being giving at this symposium are like the clay and windows, and we still have not made the wheel, pot, or house. The "house" of agromedicine cannot be built overnight, but is rather something, I feel, that will develop gradually through the enthusiasm, cooperation, support, and efforts of a great many people. I see this symposium as a door of a house that will someday come into being. It is my hope that you will all pass through this door and go on to help build the house as you see fit.

I would be delighted if this symposium serves as a new forum for the communication of information on research and education in the field of agromedicine. Hopefully it will lead to the creation of a "house" of agromedicine, even if only a very modest one.

References

Joho : Nou − Kankyo − Iryo (in Japanese) [Newsletter: Agriculture, Environment, and Medicine]pub. Kitasato University. Available at the following Kitasato University web page: http://www.kitasato‐u.ac.jp/daigaku/noui/noui.html

Chapter 3

Chiba University Center for Environment, Health, and Field Studies: Philosophy and Practice

Toyoki Kozai

Foreword : The various stresses of present-day urban living

People living in modern society, particularly urban environments with high population densities and the latest technology, face an array of emotional and environmental stress. Emotional stress can lead to such conditions as a loss of the will to live, loss of contact with others, social withdrawal, NEET (Not in Education, Employment or Training) status, and depression, while environmental stress manifests itself as increasing waste output, pollution, loss of nature, resource depletion, and such like. These phenomena are related to a breakdown in social norms, increase in crime, growing poverty, and increasingly fierce competition. In countries with declining and increasingly graying populations, moreover, rising healthcare costs and taxes further complicate the above issues. These different stresses are related to and affect each other in complex ways, which means that rather than trying to eliminate them singly, we need to take measures to identify and tackle common underlying factors.

A more detailed presentation of the contents of this article can be found at
http://www.chiba-u.jp/message/president/challenge/index.html

Keywords for the relief of stress : Oriental philosophy and culture, horticulture and plants

The various issues and stresses mentioned above are spreading worldwide, and becoming increasingly serious. Many different countries have come up with ideas and methods for resolving them—ideas and methods that can often be encapsulated by such keywords as recycling, low-impact, eco-friendly, sustainable development, resource-efficient, return to nature, slow life and slow food, finding fulfillment and a sense of community, working together, safety and peace of mind, intellectual value, "What a waste!", "Let's do something about it!", and so forth. We at our Center have added the keywords of horticulture and plants, and oriental philosophy and culture to this list.

Mission and goals

The goals of the Center were defined as follows when it was founded:

(1) Creation of an environment that ensures human health, and particularly the health of children, the aged and handicapped, and following generations

(2) Creation of a symbiotic society in which health, welfare, caregiving, education, and production are approached from a unified mind — body approach

(3) The practice of healthcare that utilizes life force and natural recuperative powers, and the practice of resource efficiency, protection of the environment, recycling of materials, cultural creation, bioproduction, and horticulture, along with the emotional satisfaction that derives from the above practices

(4) Practical research and education and training of human resources based on interaction with industry and local communities

The center's research areas

The key areas of research envisaged at the time of founding of the Center are as follows:

(1) Use of the beneficial effects of contact with plants and nature in the practice of oriental medicine and caregiving, and the practice of preventive medicine, environmental education, and horticultural therapy

(2) Universal design and use of facilities and equipment for caregiving, rehabilitation, and crop production

(3) Development of agricultural production systems that put priority on health and producer satisfaction

(4) Resource efficiency and environmental protection in a "garden city" that incorporates agriculture, horticulture, and resource recycling

(5) Cultivation, propagation, and utilization of health-promoting functional plants

(6) Development of functional plant species and resource-efficient, eco-friendly urban horticulture systems employing cutting edge technology

(7) Integration of environmental policies with welfare and caregiving policies, and environmental auditing required to achieve the above goals.

The center's organization, location, and facilities

1) Organization and location

The Center was established in April 2003 through expanding and converting the Horticulture Department's farm into a joint education and research facility. The Center is located in the Kashiwanoha district of the city of Kashiwa in Chiba Prefecture, and boasts an area of 17 ha over which the main building, lecture halls, Kashiwanoha Clinic, small-scale fields, orchards, greenhouses, processing facilities, and so forth are scattered.

The Center has 15 fulltime faculty members, 10 farm technicians, 3 fulltime and several part-time office staff members, and about 50 joint faculty members. Of the 15 fulltime faculty, 6 came from the Horticulture Department, 3 from the attached farm, 2 from the

Medical Department, 2 from the Education Department, and 1 from the Pharmacology Department. A fulltime teacher in traditional herbal medicine was recruited from Toyama Medical and Pharmaceutical University. There are probably very few research centers in Japan and even worldwide that combine such a range of disciplines.

2) Kashiwanoha Clinic

In June 2004, 14 months after the Center was launched, Kashiwanoha Clinic, a single-storey, 500 m^2 clinic located in verdant surroundings within the Center's campus opened as a clinic specializing in treatment based on oriental medicine. Equipped with examination rooms, waiting room, pharmacy, horticultural therapy room, bathroom, and other facilities, the clinic also boasts a medicinal herb and horticultural therapy garden outside on the south side of the horticultural therapy room, and pursues research in the relationship between human health and the environment, with an emphasis on prevention and caregiving.

3) Chemical-free housing and town model (sick building syndrome treatment facility)

To contribute to future community design, a model town will be built in 2006 within the Center's campus to conduct research into environmental improvement-based preventive medicine related to sick building syndrome.

An Environmental Medicine Department will be established within the Kashiwanoha Clinic to conduct (1) research in the treatment of sick building syndrome based on various housing models (detached house, tenement, condominium); (2) research into the reduction of chemicals used in housing; (3) research on evaluating and reducing the chemical contents of building materials, furniture, household appliances, cars, etc.; and (4) training of specialists.

Examples of ongoing research

1) Integration of oriental medicine with horticultural therapy

Horticultural therapy aims to improve both physical and emotional wellbeing through horticulture. Horticulture is known from experience not only to improve bodily functions without undue exertion, but also to nourish the soul and encourage positive attitudes. One of the Center's research themes is the consolidation of this hands-on knowledge jointly by doctors, pharmacologists, nurses, horticulturalists, and educators to create a systematic therapy method. We have also already launched research on a program to improve both physical and mental functions through combining horticultural therapy with oriental medicine to generate synergistic effects. The above research projects commenced in the summer of 2004, and the results of the Center's first phase of research were published in August 2005 (Noda, 2005). We plan in future to complement and combine the above research with aromatherapy, acupuncture, forest bathing, Western medicine, and so forth.

2) LOHAS urban design

Right from the start, we aimed to design the Center campus and influence the development of its surroundings in a way that reflects the ideals and mission of the Center. To ensure that the Kashiwanoha campus is a fitting symbol of a new era of emotional fulfillment and care for the environment, we decided to (1) make full use of Chiba University's collective capabilities, (2) consider the role of the campus in relation to the surrounding community, and (3) place top priority on health and environment, while (4) also considering income generation.

We also decided to employ LOHAS (Lifestyles of Health and Sustainability) concepts in designing the surroundings of the Kashiwanoha Campus Station. LOHAS consumers seek (1) eco-friendly lifestyles, (2) the creation of a sustainable economy, (3) healthy lives with a priority on prevention of illness rather than drug-based healthcare, and (4) to put priority on self-expression.

We are planning to line the road of about 1 km by 2008, leading from the station, through the campus, and on to Kashiwanoha Park with double-flowered cherry trees, and proposals have been submitted for facilities in the vicinity of the station such as a Yakuzen (cooking with traditional Chinese herbal remedies) health food restaurant and shop, herbal medicine pharmacy, community farm, acupuncture clinic, organic greengrocer, and nursing home and other care facilities that incorporate horticultural therapy.

3) Closed-system herbal medicine production

With the graying of the populations of developed countries, the rapid economic growth of Asian countries (particularly China), and growing interest in healthy living worldwide, the demand for herbal remedies is rising. However, owing to the overharvesting of wild-growing medicinal plants, stocks have declined significantly. The need to protect the natural environment and boost the production of ingredients for herbal remedies is expected to drive the development of cultivation systems. The development of cultivation technologies for boosting the growth and increasing the active ingredient content of medicinal plants, and the cultivation and patenting of genetically superior variants in terms of growth rates, active ingredient content, and tolerance to disease are going to be crucial to the development of the herbal remedy health industry.

In view of the above circumstances, the cultivation of medicinal plants in specially designed facilities located in key consumer nations is likely to become a major trend. As one such possibility, the Center is looking into the cultivation of medicinal plants based on a closed system. The features of such systems are: (1) enclosure in insulated walls that shut out light; (2) use of artificial light only; (3) minimum ventilation; and (4) minimum import and export of materials into and out of the system
(Kozai et al. ; http://phdsamj.ac.affrc.go.jp/topic/5_2.html;

http://www.taiyo-kogyo.co.jp/naeterasu.html).

Closed — system cultivation offers the following advantages: (1) cultivation is unaffected by external climatic conditions; (2) resource-efficient and eco-friendly, requiring less water, fertilizer, and CO_2 enrichment; (3) ability to shut out pests makes pesticides unnecessary; (4) equipment and operating costs are cheaper per production unit than greenhouses; (5) labor- and space-efficient, requiring only 1/10th the space of greenhouses; (6) maintenance of an ideal environment enables fast growth and production of higher concentrations of active ingredients; (7) high-quality produce owing to exclusion of insects, germs, pesticide residues, dust, and other impurities; and (8) vegetative propagation from cuttings of the best stock enables the cultivation of genetically uniform, high-quality crops. These merits have been demonstrated so far by research involving closed-system cultivation of St. John's wort (*Hypericum perforatum* L.) (Mosaleeyanon et al., 2005), Chinese licorice (*Glycyrrhiza uralensis*) and other medicinal plants.

Community engagement activities

1) Public lectures on environment and health
(http://www.h.chiba-u.jp/center/event/event.htm)

To forge closer relationships with the local community, the Center has, since its Seeds Hall lecture room was completed in February 2004, held regular "Environment and Health" lectures mainly on weekends (organized by fulltime Center faculty members Katsuji Noda and Satoru Tsukagoshi,). Most lectures were given by fulltime Center faculty members, but depending on subject, joint faculty members or outside experts have also been invited to give lectures. By autumn 2005, 26 lectures had been held.

2) Environment and health business forums

We held 8 "Environment and Health Business Forums" from May to December 2004 mainly for the purpose of exchanging information and views with representatives of private enterprise and local government. The first 3 forums were devoted to explanation of aims and to self-introductions by participants. The ensuing 5 forums each featured the presentation of a specific theme, followed by open discussion. The themes discussed were: (1) Current status and outlook for healthcare based on oriental medicine: the practice of medical treatment in tune with nature; (2) Closed-system plant cultivation; (3) Kashiwanoha Campus and urban design; (4) Horticultural therapy: overview and initiatives; (5) Development of simplified methods for measuring trace quantities of chemical substances. Each forum drew a participation of 70-150.

3) Senryoku-kai volunteer group

Senryoku-kai (literally "One Thousand Greens Association") volunteers donate their time to a wide range of activities, including organization of the almost monthly Environment and Health public lectures, cultivation of plants for the horticultural therapy garden, and help in harvesting crops. Volunteers also participate as subjects in the Center's horticultural therapy research experiments. Almost all Senryoku-kai members are very active types who love getting their hands dirty tending crops, and they derive real pleasure from participating in activities of the Center. We feel that Senryoku-kai could become an ideal model for community engagement and volunteer activities to support public research and educational institutions.

Closing comment

The research, education, and community engagement activities being carried out at Chiba University Center for Environment, Health, and Field Studies are not by any measure cutting edge endeavors. In fact, they could almost be seen as areas left behind in the rush, but we feel that we are tackling issues of vital importance to modern society, and we put priority on the steady but sure resolution of problems rather than on speed and efficiency. It is our fervent wish that the efforts of the Center will spread throughout the world and contribute to welfare in the 21st century, a century that we position as an era of emphasis on environment and emotional fulfillment.

Chapter 4

Agromedicine from the Perspective of Medicine

Yoshiharu Aizawa

Foreword

From my 3rd to 6th year as a medical student, I belonged to a society in the med school dedicated to providing healthcare services to communities lacking a fulltime doctor. Over the summer vacation, we would enlist the services of doctors in the university hospital to carry out health checkups, while in the winter break, teams of doctors plus med students would go around calling in on every household in a community to check on any health problems. I helped provide such services in my 3rd year in the villages of Okura and Sakegawa near Shinjo in Yamagata Prefecture, and in my 4th, 5th, and 6th years in the town of Kawanishi, which is now part of the city of Tokamachi in Niigata Prefecture, a region noted for high snowfall. This was 1968–1971, a time when Japan's economy was red hot, and almost all young folk in farming households apart from the eldest son headed for the cities, and even most of those who remained would leave to do seasonal work elsewhere. It was a time when Japan was industrializing rapidly, and farming held little appeal as a future. Owing to excessive government protection, the agricultural sector failed to modernize at the pace it should have done as a genuine food production industry, and the country's food self-sufficiency rate accordingly dropped to 40% in terms of energy. In this session, I would like to consider agromedicine from the perspective of medicine, looking back on my experiences of that time. And speaking in my present capacity as a specialist in community medicine, I would also like to propose the establishment of dietary science as a discipline in Japan.

Farming population

In 1975, 11% of Japan's population worked in the agriculture sector, but since then the country's industrial structure has changed dramatically, causing the farming population to drop to 4% (3.82 million people). Migration to the cities is said to have diminished in recent years, but those remaining in rural communities still tend to be elderly, the percentage of people of 65 years and over in such communities in 2000 standing at 28.6%, compared with a national average of 17%. A full 52.9% of those who give farming as their main occupation are 65 and over. Considering that the equivalent for the UK and France is 7.8% and 3.9%

respectively, Japan's farming population is conspicuously elderly. In my student days, why people would choose to leave country villages for the cities with their foul air and water, and what the future held for Japanese agriculture were subjects we discussed endlessly, but we could not help but be pessimistic about the outlook for both agriculture and rural communities. Our perception at the time was that young people probably headed to the cities in search of a modern and prosperous lifestyle, but now when I think about it, it could also be seen as an aspect of reproduction, since it would be only natural for young people seeking marriage partners to migrate to the cities where they would have much greater choice. Young people are also attracted to industries with good growth prospects, and those industries also flourish as a result of the intake of young people.

With the enactment of the New Basic Law on Food, Agriculture, and Rural Areas in 1999, Japan changed direction from protection of farmers to policies for promoting sustainable agriculture from the perspective of the public at large. The new law positioned agriculture as a food production industry and implemented policies aimed at improving the lives of food producers through developing the industry itself. In 1999, the farming population stood at 3.24 million households, of which about 480,000 were fulltime farmers, but with the change in policy, by 2010, farming households are expected to drop to 2.3−2.7−million, with fulltime farmers practicing efficient farm management that generates a steady income numbering 330,000−370,000, and agricultural co−ops and other incorporated food production bodies numbering 30,000−40,000.

The importance of dietary education

It goes without saying that the vitality of a society depends on the vitality of the people that make up that society, and it is physical and mental health that underpins the vitality of individuals.

To maintain physical and mental wellbeing over their lifetimes, people need to follow appropriate dietary, exercise, and sleeping habits. Dietary habits in particular tend to be heavily influenced by the kind of dietary regimen that people are exposed to as children. Good dietary habits in the formative years are also essential to the healthy development of both body and mind. In short, the practice of good dietary habits as children leads to a more physically and mentally healthy adult population and serves as a foundation for enhancing the vitality of society as a whole. In recent years, changes in the social environment as it affects dietary habits have resulted in unbalanced nutritional intake, increasing obesity, unhealthily skinny physiques and other worrying outcomes, and poor dietary habits are also thought to be closely linked to lifestyle−related diseases. As such, the promotion of correct dietary habits could be regarded as a matter of national concern. In the light of such circumstances, the government passed a Basic Law on Nutritional Education on June 17, 2005. While "skills education" is a particularly important component of the education of medical specialists, dietary education needs to be positioned as a 4th fundamental component of education in general, joining the 3 pillars of intellectual, moral, and physical education. Dietary education should be focused on the nurturing of people capable of practicing healthy dietary habits through furnishing them with practical knowledge about nutrition and diet and the ability to make appropriate choices regarding the food they eat. In addition to diet and nutrition, this education should also cover such subjects as food safety and cultural aspects of food and diet.

Appropriate nutritional intake plays an important role in the prevention and treatment of lifestyle-related diseases, and agriculture accordingly also has a major role to play in the everyday life of the general public. Snack foods and instant foods with their high fat and salt content are feared to be promoting unbalanced nutritional intake. Moreover, food is important not only for its nutritional value, but also as a cultural component that enriches everyday life. While the development of health foods, non-perishable foods, and therapeutic diets is important, it is also just as important to develop the means to evaluate the safety and efficacy of such foods and diets. Efficacy in particular is not yet being adequately assessed. Dietary education itself also needs to be scientifically evaluated for its educational content. Traditional food science too needs to be rethought and reorganized as dietary science — a discipline that covers all aspects of food and diet, rather than just foodstuffs.

Towards the integration of medicine with agriculture

By promoting cooperation and interactions between urban and rural communities, we can cultivate trust and build personal ties between food consumers and producers. This would hopefully boost consumer awareness and encourage them to take a deeper interest in food and diet, while also providing them with the peace of mind that comes with being able put familiar faces to the vegetables and other produce they consume. Such consumers would also hopefully be far less inclined to waste food, thus leading to the more effective use of food resources. All of this may in turn hopefully inject new life into rural economies in a way that is also environment-friendly.

Occupied as they are in the promotion of health, medical professionals are ideally positioned to spread the word at every opportunity about the wisdom of following good dietary habits. It is the duty of medical science to scientifically judge the benefits of dietary education and develop the required educational methodology. Equipped as it is with the resources for teaching both agriculture and medicine, Kitasato University should put those resources to good use to promote the development of dietary science and education.

Scholarship has tended to branch into many different fields as it has advanced, and such specialization often makes it difficult for any single field to independently come up with solutions to real-world problems. It is for such reasons that we are seeing interdisciplinary cooperation and joint research by different fields, but because such collaboration requires considerable energy to bring about, it rarely works unless goals are first very clearly defined. To enhance the output from agromedicine, I feel it is necessary to cultivate links between agriculture and medicine by setting concrete goals such as those listed below. I also feel that it is necessary to establish an Agromedicine Center at Kitasato University under the authority of the president to promote educational and research activities.

1) Research

a) Development of functional, non-perishable, and other types of foods, and scientific evaluation of their nutritional efficacy

b) Fact-finding surveys of dietary habits and identification of areas requiring improvement: evaluation of the health implications of eating alone, eating out, snacking, skipping breakfast, and so forth.

c) The relationship between dietary habits and mental health

d) Concrete recommendations for improving dietary habits: display of energy, salt, and nutrient content on menus

e) Prevention of lifestyle-related diseases through correct eating, development of treatments

2) Education

a) Practical, hands-on education for students in dietary education: establishment of infrastructure, interactive education between Schools of Agriculture and Medicine.

b) Development of dietary education teaching methods: creation of materials and curricula, development of software

c) Application of agromedicine to student mental health: early intervention for students having adjustment problems

d) Volunteer activities for the local farming community: participation in agriculture in the Sagamihara area agriculture

e) Acquisition of agro-ecological concepts: growing out of the mass-production, mass-consumption model

Closing remarks

Conducting the above agromedicine research and educational activities will, in my mind, require the establishment of an Agromedicine Center (provisional name) staffed by a collection of specialists in food science, nutrition, agriculture, immunology, and other disciplines. Securing field sites for the hands-on education of students and surveys and analyses is another important requirement.

Chapter 5

Food, Agriculture, and the Environment

Isoya Shinji

Negative aspects of the industrialization of agriculture

Following the 1992 Rio Earth Summit, Japan passed the Basic Environment Law in 1993, the Basic Law on Food, Agriculture and Rural Areas in 1999, the Basic Law for Establishing the Recycling-Based Society in 2000, the Basic Law for Forests and Forestry in 2001, the Law for the Promotion of Nature Restoration in 2002, the Law for Enhancing Motivation on Environmental Conservation and Promoting of Environmental Education in 2003, the Three Laws on Scenery and Greenery in 2004, and the Basic Law on Nutritional Education (a member bill) in June 2005. Diet is fundamental to health, and so it makes perfect sense to legislate for correct eating habits, but this list of laws testifies too to how the way we eat has been distorted by our runaway modern civilization.

By "runaway modern civilization", I mean the way that all of our economic behavior tends these days to be geared towards greater and greater efficiency.

In our eagerness to boost efficiency, productivity, and "the economy", we are even ready to sacrifice human life and health, and the health and sustainability of our environment. Insofar as it ignores the biodiversity and conservation of the ecosystem on which our own survival depends, this blinkered pursuit of economic efficiency at all costs — what I call the "pursuit of partial efficiency" — is indeed a runaway philosophy that has grave implications for all of us.

This pursuit of partial efficiency could, put in very plain terms, be seen as the outcome of modern scientific method, and more specifically, an engineering approach. Imagine, for example, using closed-system machinery to create a production system geared to achieving the greatest efficiency as determined experimentally under certain fixed conditions. One such example would be the chemical industry, which has boosted its productivity and competitiveness through reducing effluent treatment costs by passing the buck onto the economy at large in the form of pollution and environmental degradation. Regrettably, modern agriculture is of course guilty of committing the same errors.

It is said that modern agriculture began with the willow cultivation experiments of Flemish scientist Jan Baptista van-Helmont (1577-1644), but it really took off with the research and development of chemical fertilizers, synthesized agrochemicals, and mechanization carried out from the 19th century onwards. Spurred by the cheap energy that the discovery of oil produced, agriculture industrialized at a spanking pace, eventually giving rise to the concerns that Rachel Carson articulated in Silent Spring.

In recent years, much is being said about the multifarious functions of agriculture, and particularly the role that agriculture and forestry can play in environmental protection, but as long as we continue to rely on the same partial efficiency approach to agricultural R&D, we will not be able to come up with any real solution to the environmental problems we face.

The multifunctionality of agriculture

Human survival is unthinkable without economic activities, and so economic pragmatism is essential, but the need to preserve the sustainability of the global ecosystem and human health (which also depends on the maintenance of a dynamic balance between all parts of the body) demands that we urgently adopt an "overall efficiency" approach.

Tokyo University of Agriculture's "Eco-Eco Agriculture" Academic Frontier Project (Matsuda; Fujiki et al.) to develop new agricultural methods that combine ecology with economy could be seen as a concrete initiative aimed at implementing such an overall efficiency approach.

Japan's agricultural policy ever since the Agricultural Basic Law of 1961 has been devoted in the main to improving the economic productivity of agriculture solely as an industry. I, however, have long advocated considering (1) farmland (space), (2) farmers (human capital), (3) farming households (basic units for handing down culture, training successors, and sustaining agriculture), and (4) farming villages (political, economic and social units; local communities for the communication of history and culture) as an aggregation of assets that could be referred to collectively as agriculture or farming, and which include not only production assets, but also social, cultural, and environmental assets. It is because I considered taking an overall view to agriculture as being important that I also ensured that the content of the 2nd Basic Plan for Food, Agriculture and Rural Areas was peppered with references to rural landscape.

As the only industries found in all parts of Japan, farming and forestry are highly important from the perspective of environmental protection. As such, we need to formulate policies for the establishment of practices that make optimum use of all of their multifaceted functions.

Incidentally, according to the calculations of the Science Council of Japan (2001), the value of agriculture's many-sided functions works out at about 8.3 trillion yen a year, and that of forests at 70.3 trillion yen. I personally feel that these estimates are too low, since the annual turnover of Toyota, a single company is 16 trillion yen, which is over double the yearly value of agriculture mentioned above.

The reason that the EU's Deral says "Agriculture is more important than money, since agriculture is culture" is much the same as my reasons for inventing my own definition of agriculture after the Japanese for farmer (*hyakusho*, "one hundred names") as "a way of life that requires and makes the most of many different capabilities and is still essential in this age of specialization and industrialization." I feel that greenery and agriculture are truly entities of immeasurable, limitless value.

The naturalness and wholeness of "man and environment"

Modern agriculture's embrace of specialization and industrialization has confused and degraded the Earth and mankind, which are fundamentally holistic in nature.

We need now, more than ever before, to get back to agriculture as it used to be practiced — as a sustainable and organic means of producing food in tune with the natural environment — and also to rediscover ourselves, humanity, as a totality.

I use the expression "farming landscape" and hyakusho, the Japanese word for farmer, in the hope that we can indeed restore the multifariousness and integrity of both agriculture and humanity as parts of a larger whole.

In this sense too, I think it is of vital importance to look at food, agriculture, and the environment from a holistic viewpoint, or if you will, as a single "landscape".

This approach of mine is in fact very much akin to that of a landscape artist, grounded as it is in the combination of "land" with an overall, holistic view ("scape"). It is from this perspective that I argued in a book titled "The Age of Agriculture" (pub. Gakugei Shuppansha, 2003) that if the 20th century was an age in which farming villages became urbanized, the 21st century is an age in which we should promote the ruralization of cities. As a negative legacy of the 20th century, our huge and packed cities have given rise to an increasing array of urban problems and pathologies — crime, murder, mental illness, technostress, neurosis, loss of will to live and other symptoms of precocious senility. As an antidote to these pathologies, I am arguing, albeit relatively, for the urgent need for a switch to lifestyles at the opposite pole from our huge, man-made cities, to lifestyles imbued with at least an element of agriculture and which offer opportunities for experiencing nature, countryside, life in farming villages, and so forth.

We human beings are, moreover, "wholes" made up, in the Western phrase, of a harmonious combination of body, mind, and spirit, and I think that we instinctively resent the way the division of labor under the differentiation and specialization of modern society has made it difficult for us to maintain this unity of body, mind, and spirit. I feel that we are instinctively seeking to recover the "wholeness" of the hyakusho lifestyle, and that this is fueling the rising interest in cultivation, horticulture, and "the country life" as a kind of shortcut to the recovery of wholeness. We are, moreover, first and foremost living creatures, and no matter how far civilization advances, it would be inconceivable without the existence of the natural environment. As such, I am arguing in absolute terms for the following types of coexistence: (1) coexistence with other living creatures and nature; (2) environmental coexistence with resources and energy; and (3) regional coexistence between city and farming village, developed and developing countries, and so forth.

Agriculture as the link between food and environment

In the past, people leading such holistic farming lives cultivated healthy food, and maintained a more healthy symbiotic relationship with the environment. However, the advance of industrial civilization has in 100 years boosted the global economy 20-fold and energy consumption 25-fold, despite the fact that the global population has only quadrupled over the same period. The growth of industry has created societies of mass production, mass distribution, and mass consumption, together with horrendous waste, giving rise to serious environmental problems that put the survival of humanity itself at risk.

To escape from this threat, we need to rethink our pursuit of specialization and partial efficiency, and reinstate concepts and practices grounded in holistic principles. The word "holistic" is, by the way, derived from the Greek word "holos", as are such words as "whole", "heal", "health", and "holy".

In academia too, we have seen increasing specialization. There are currently over 50 different societies registered with the Science Council of Japan under the Association of Japanese Agricultural Scientific Societies. This number testifies to the trend towards ever greater specialization in academia and university education.

With the aim of stemming this tide even if only slightly, in November 2004 I and others founded the Society of Practical Integrated Agricultural Sciences (president: Eiji Yamagiwa, vice presidents: Isoya Shinji, Katsuyuki Minami). While we recognize the need for research in specialized areas, we founded the Society because we felt it important to launch a movement consciously directed at the integration of various specialized fields, and at the practical application of its findings for the betterment of society. I chose the name Food, Agriculture & Environment for our journal (#1 issued in April 2005, #2 in December 2005) out of a desire to drive home the point that agriculture is the only link between food and environment, and their only common means of support.

Agriculture-oriented lifestyles

Figure 1 is my vision for (1) rekindling contact with agriculture and reestablishing the link between food and environment in the lives of city dwellers suffering from the illusion that distancing themselves from the soil and greenery represents progress and (2) conserving and rehabilitating what from local to national level is a natural environment made up largely of secondary nature that requires human intervention.

The government already recognizes the importance of this kind of course of action, and in its 2nd Basic Plan for Food, Agriculture, and Rural Areas being implemented since 2005, articulates its intention to "promote cooperation and interaction between cities and rural areas, and the participation of diverse parties in such initiatives". This policy also stems from a realization that Japan's population engaged in farming, fishing and forestry, now numbering under 5% of the total population, is too small to shoulder responsibility for the conservation of the 67% of Japan's territory covered in forest and the 13% currently comprising agricultural land.

In view of such circumstances, I see a need to proactively pursue such policies as the establishment of a special official category of part-time farmers encompassing the whole population, or the promotion of "21st century nature-oriented lifestyles". The Basic Plan mentioned above presents a model composed of the following 4 lifestyle levels: (1) permanent migration to rural areas, new fulltime farmers; (2) semi-permanent residence (dual urban/rural lifestyle); (3) recreational agriculture (participatory and volunteer rural village stays, urban community farms, etc.); (4) urban-rural interactions (e.g. agrotourism, sister city arrangements, etc.).

I feel that unless such proposals for new, agriculture-oriented lifestyles are implemented as national projects or popular movements, there is little chance of creating a physically and mentally healthier populace or society through the integration of food, agriculture and environment.

To achieve such ends, I want to stress again the need to base our efforts on a clear understanding of the seriousness of modern urban pathologies and the special characteristics, value, and efficacy of agriculture and rural areas.

Environmental awareness for environmental welfare

Many people are of course unable to pursue rural life owing to their occupations, but the kind of nature-oriented lifestyle proposed in Figure 1 is more a matter of attitudes, something that can be practiced by anyone anywhere.

Architectural critic Noboru Kawazoe once said that almost all Japanese have a shared liking for festivals and horticulture, and it is with the same sentiment that I and others in 2001 founded an NPO named Japanese Society for the Promotion of Horticultural Welfare dedicated to encouraging people to discover the joy that cultivating flowers and vegetables can bring. As of January 2006, the Society has about 1,500 members, and 1,220 novice horticultural welfare specialists. Each year 1,200 people enroll in courses, with about 800 of those taking the novice horticultural welfare specialist exam. These specialists are now making a major contribution to society in such fields as welfare, medicine, and town planning, and have already held 6 national conferences. Through cultivating flowers and vegetables, people can join others outdoors in tilling the earth and enjoying the experience of nurturing life to cultivate safe food and contribute to the preservation of agriculture and environment.

Many people these days wish to contribute in some way to the solution of environmental problems. I refer to such people as "environmental citizens" or "environmental students", and have put together a number of introductory guides for them.

I decided to get involved in such activities out of a feeling that "economic welfare" — the use of money to make people happy that has been the norm up to now — had reached its limits and that what was needed now was "environmental welfare" (the creation of happiness through joining other like-minded people in a pleasant environment in activities that benefit everyone).

I felt that educating students with such inclinations (environmental students) and creating a network of people with such practical capabilities (environmental citizens) was a worthwhile endeavor.

In Table 1 I have listed the various reforms and policies for contributing to society that I implemented during my term as president of Tokyo University of Agriculture. The new Departments of Biotherapy and Aqua Biosciences opened in April 2006 are recent outcomes of such initiatives. Horticultural and animal therapy are perhaps the ultimate examples of biotherapy, but I felt the need to create a broader foundation — bio — welfare, if you will — on which to base such therapies. In Figure 2, Eisuke Matsuo (2004) shows the relationships between these components which, just as Hajime Orimo (2003) does in Figure 3 to show the need for integrated medicine, demonstrates too the fact that solving any problems these days ultimately requires a staged approach and the complementary use of various technologies and disciplines.

I am also involved in a number of other NPOs and initiatives. These include, at the overlap between food and agriculture, the Good Foodstuffs Promotion Association, and at the overlap between environment and agriculture, Fingers of Green (footpaths, woodland and farmland conservation), the Rural Nature Rehabilitation Contest (sponsored by the Ministry of Agriculture, Forestry and Fisheries and the Ministry of the Environment), the Rural Area Scenery Supporters Club (sponsored by the Ministry of Agriculture, Forestry and Fisheries), and the Beautiful Land Creation Association. All of these are holistic grassroots initiatives aimed at rehabilitating the relationship between humanity, food, agriculture, and the environment.

I am very fond of the word "reap" for the way it hints that we should all take some kind of action. In the words of American clergyman G.D. Boardman:

> **Sow a thought, and you reap an act ;**
> **sow an act, and you reap a habit ; sow a habit, and you reap a character ;**
> **sow a character, and you reap a destiny.**

It is not just medicine and food that share a common foundation. Medicine, agriculture, food, environment, and health are all fundamentally connected through the living creature that is man. The Japanese expression umashi kuni ("beautiful land") refers not only to visual beauty, but also to "amenity" in the sense of historical and natural goodness (the root of the word amenity, in turn, is thought to be related to amare [love]). Beauty derives from the harmonious integration of all the parts into a single whole.

Chapter 6

Combining Oriental Medicine with Horticultural Therapy

Toshiaki Kita

Foreword

In recent years, we are seeing a major paradigm shift in the concept of health, owing largely to the inadequacy of the traditional paradigm of health and illness as diametrically opposed elements in resolving the medical issues faced by modern society. To promote preventive medicine (primary prevention), we need a new paradigm that recognizes a state that is neither healthy nor ill, but somewhere in between (known as mibyo ("un‑ill") in oriental medicine), a paradigm for promoting health such as Antonovsky's theory of salutogenesis (the creation of health). I feel that oriental medicine and horticultural therapy are special for the way they serve as approaches not only to the causes of illness, but also to the creation of health.

Western medicine: elucidating the cause of illnes

Modern Western medicine has in general been based on a reductionist, mechanical theory of the origin of disease. For example, the 2005 Nobel Prize for Medicine was awarded to Drs. Barry Marshall and Robin Warren for "their discovery [in 1982] of the bacterium *Helicobacter* pylori and its role in gastritis and peptic ulcer disease". The researchers were recognized for proving the very close link between *H. pylori* and stomach inflammation and ulceration of the stomach and duodenum, and elimination of *H. pylori* is now the most common way of treating recurrent peptic ulcers. This is a typical example of the way Western medicine has advanced through elucidating the mechanism behind the occurrence of diseases in terms of simple linear, deterministic relationships between etiology and pathology (cause and effect), and then inventing new treatments accordingly. It goes without saying that this etiological approach will continue to play an important role in medicine.

Oriental medicine : normalizing health creation (salutogenesis)

However, despite the fact that 80% of Japanese over the age of 40 are infected with *H. pylori*, only 3%-5% ever suffer peptic ulcers, and most carriers enjoy a relationship of healthy coexistence with *H. pylori*. In other words, if the "salutogenic" mechanisms that the human body is equipped with to maintain health are working properly, they can prevent disease and cure it naturally when it does occur. Oriental medicine aims both to remove the factors causing disease and to simultaneously invigorate the natural healing powers we all possess, thereby normalizing salutogenesis. For example, a doctor of oriental medicine might diagnose peptic ulcers as evidence of a decline in natural healing powers and prescribe Ninjinto, a herbal remedy, or on the other hand, diagnose ulceration as evidence of a rise in factors causing disease, and prescribe Orengedokuto, a different herbal remedy.

The role of oriental medicine in modern healthcare

Many of the major healthcare issues faced by present-day society stem from a rapid increase in the elderly population, increasing stress, and a shift in the disease landscape towards lifestyle-related diseases. The 3 most common threats to salutogenesis too are age-related changes, stress, and lifestyle — related diseases. For example, the reason why we tend to suffer from an increasing number of afflictions as we grow older is that salutogenic mechanisms are impeded with aging. A 74-year-old man came to me complaining of increasing physical fatigue. He was suffering from diabetes and related complications and was accordingly receiving separate treatments from specialists in internal medicine, ophthalmology, orthopedics, and urology respectively for his diabetes, cataracts, osteoporosis-related back pain, and frequent urination caused by an enlarged prostate. I diagnosed him as suffering from what in oriental medicine we refer to as insufficient ki or "life force", and prescribed Hachimijiogan, a herbal remedy that had the effect of lightening his various symptoms, enabling a reduction in the amount of conventional medicines he was taking. Oriental medicine can play a major role in healthcare for the elderly for the way it boosts natural healing powers and improves quality of life (QOL).

Investigation of health status using health-related QOL indices

In general, a person whose salutogenic mechanisms are working properly will be healthy, while someone whose salutogenic mechanisms are impeded will experience a decline in health. However, there are no scientific indicators in existence for objectively rating level of health, and so I decided to use the Physical Component Summary (PCS) and Mental Component Summary (MCS) of the SF-36 v2 (Japanese language edition), a health-related QOL survey that is used widely throughout the world, as indicators to investigate how the clinical condition — from the perspective of oriental medicine — of patients attending the Kashiwanoha Clinic influenced their level of health, and the effect of herbal treatments on health level. Results showed that if the oriental medicine-based clinical condition is not serious, both PCS and MCS show little decline, but if it is serious, both PCS and MSC decline significantly. Results also showed that even relatively short periods of 4-8 weeks of herbal remedy treatment improved both PCS and MCS significantly. Oriental medicine (herbal treatments) not only improves health-related QOL that has declined owing to impediments to salutogenic mechanisms, but also represents a *mibyo* ("un-ill") concept-based solution to the increased risk of contracting disease.

The many different benefits of horticultural therapy

At the Chiba University Institute of Environment, Health, and Field Sciences, in addition to the core practice of oriental medicine whose characteristics I described earlier, we are also looking into various ways of putting our verdant surroundings to good use, particularly for horticultural therapy. Horticultural therapy involves the addition of various therapeutic processes — in short, processes for improving condition, restoring functions, and raising QOL — to the practice of horticulture by people (subjects of healthcare and welfare). Horticultural therapy brings many different benefits, including (1) emotional solace derived from cultivating plants and experiencing the workings of life and the rhythms of nature and the seasons; (2) benefits derived from the diverse pleasures of growing plants, harvesting, eating, creation, and decoration; (3) benefits derived from exercising the body and developing manual dexterity; and (4) benefits derived from sharing plant growth and conversing and empathizing with others. Horticultural therapy is thought to boost natural healing powers through different channels from oriental medicine (herbal remedies), and as such, combining them both can be expected to generate synergistic effects.

Horticultural therapy and sense of coherence

Horticultural therapy might work by bolstering the sense of coherence (SOC) that is said to play the most important role in salutogenesis, and this is perhaps an outstanding attribute that is absent from oriental medicine. According to Antonovsky, SOC is a feeling of confidence in one's ability not only to understand and deal with the stressful demands in one's life, but also to regard these demands as worthy challenges.

It seems likely that by participating in horticultural activities and observing how plants grow according to the laws of nature, and by cooperating with others to successfully carry out horticultural tasks and experiencing how plants respond to the care given to them, one can regain one's sense of coherence. If you actually observe people participating in horticultural therapy, you would need no convincing by others that horticultural work being carried out with obvious enjoyment, by people on their own or in groups, is clearly contributing to better health. (However, finding objective proof and scientifically elucidating the mechanisms involved is far from easy.)

Closing comments

At the Chiba University Institute of Environment, Health, and Field Sciences, we are seeking to contribute to the health of the Japanese populace first and foremost through the practice of oriental medicine and horticultural therapy, but also through yakuzen therapy (cooking with traditional Chinese herbal remedies), environmental health education, and other means that utilize the Center's horticultural and health resources. We have still only just started to combine oriental medicine with horticultural therapy, but I hope that this talk will have been of some use to future contemplation of the subject of agromedicine.

Chapter 7

Human Health and Functional Foods

Takafumi Kasumi

Foreword

Health and safety are probably the two aspects of foodstuffs that most interest people in the present age. Everyone hopes for perennial youth and long life, and looking back into the past, people have over time practiced all sorts of customs related to food that are almost akin to religious belief.

The annual cost of national health care in Japan currently stands at about 31 trillion yen, and if it continues to grow at the present rate, it is expected to climb to 50 trillion yen in the near future. Such a rise signals the inevitable collapse of the system itself, and all sorts of reforms including steeper charges for senior citizen healthcare and higher patient co-payment rates are being implemented. The main reason for the steep rise in healthcare is the increase in lifestyle-related diseases, such as cancer, cardiac disease, and strokes. This has given rise to a shift in the focus of healthcare from treatment to prevention and, as an aspect of this shift, expectations are rising for the role that food and diet can play in the disease prevention and health promotion.

Food function and functional foods

Food function and functional foods are concepts advocated in the (then) Ministry of Education, Culture, Sports, Science, and Technology's "Systematic Analysis and a View of Food Function" research project (1984-1986), which proposed that in addition to nutrition (primary function) and palatability (secondary function), food performed tertiary functions such as bodily defense and regulation of physical condition. Proposing as it did that the daily intake of food (ingredients) not only supplied vital nutrients and maintained health, but also played an active part in the prevention and treatment of disease.

The project represented a landmark that attracted the attention of research institutes and the food, agriculture, and fishing industries as well as consumers. The findings of the project were also introduced overseas in the journal *Nature* (1993), prompting the creation of infrastructures for the study of food function in the West, where functional food science was established as a distinct discipline, and vigorous efforts were launched to apply research findings to the development of functional foods. Food function and functional foods became terms with global currency, but the latter is bound by no legal definitions or restrictions and is regarded internationally more as an idiom used by food-related industries. As such, the treatment of functional foods in Japan and the West in terms of nutritional value and health maintenance functions is not necessarily of the same level.

Designated health foods and dietary/food supplements

Japan has a system for permitting vendors to make health claims with respect to foods or food ingredients that have been proven to perform physiological functions and judged from human intervention trials to be suitable for the purposes of preserving human health. There are now over 500 of these "designated health foods". The term "functional foods" is often used in Japan in a narrow sense to refer to these designated health foods, whereas in USA it is used in a broader sense to include food and beverages with reduced fat, carbohydrates, cholesterol, and so forth, with functional foods in the Japanese sense being referred to as "dietary supplements". Most of these do not take the form of foods as such, and are positioned somewhere between foods and drugs, containing vitamins, minerals, amino acids, herbs, plant extracts and so forth, and are targeted to semi-healthy conditions that are a mix of over- and under-nutrition. In Europe, the same products minus herbs and other ingredients are marketed as "food supplements".

Japan's health food (or so-called health food) market is currently worth about 1.2 trillion yen annually, with designated health foods accounting for about half — 600 billion yen. In USA, dietary supplements and functional foods worth \$20 billion are sold annually, while in Europe, functional foods worth 10 billion euros (\$12 billion) are sold annually.

Studies on the use of foods for primary prevention of lifestyle-related diseases

The following studies have been carried out on the use of foods to prevent lifestyle-related diseases:

1) **Prevention of cancer**
   - Reduction of the carcinogenic properties of carcinogens (polyphenols, carotenoids, dietary fiber, broccoli sulforaphane)
   - Destruction of cancer cells through stimulating the immune system (polysaccharides, including seaweed fucoidan and mushroom or barley β-glucan)
   - Induction of differentiation and apoptosis of cancer cells (flavonoids such as isoflavone, quercetin, phloretin, and proanthocyanidin)
   - Induction of apoptosis (flavonoids)

2) **Prevention of atherosclerosis**
   - Prevention of hyperlipidemia (polyunsaturated fatty acids DHA and EPA, α-linoleic acid, barley β-glucan, seaweed alginic acid, fruit pectin, soy protein, soy saponin, green tea catechin, etc.)

- Prevention of oxidation of low-density lipoproteins (vitamins C and E, carotenoids, green tea catechin, red grape skin constituents anthocyanin and resveratrol, etc.)
- Prevention of homocysteinemia (vitamins B6 and B12, folic acid)

3) Prevention of hypertension
- Prevention through inhibition of angiotensin converting enzyme (ACE) (casein tripeptide, sardine muscle dipeptide, soy tripeptide, etc.)
- Search for hypertension-preventing substances using spontaneously hypertensive rats (SHR) (GABA tea [green tea containing GABA (γ-aminobutyric acid)], germinated brown rice, fermented rice bran extract, chitosan, vitamin C, α-lipoic acid, etc.)

4) Prevention of diabetes
- Suppression of acute hyperglycemia (dietary fiber, gymnemic acid, arbutin, phloretin, incretin inhibiting peptide, etc.)
- Suppression of glycation of proteins etc. (flavonoids, pyruvic acid, chelate compounds, etc.)

5) Prevention of aging and dementia
- (polyunsaturated fatty acids DHA and EPA, folic acid, vitamins B6 and B12, α-lipoic acid, egg yolk lecithin, arachidonic acid, etc.)

6) Prevention of allergies
- (lactic acid bacteria and bifidobacteria, citrus fruit and green tea flavonoids, β-glucan, oligosaccharides, etc.)

Nutrigenomics and personalized foods

Important aspects of food function research and the development of functional foods include
- building a database of functional food factors in relation to the maintenance and improvement of health,
- conducting molecular level genome-based analyses (nutrigenomics), and
- developing personalized foods.

1) Creation of a food factor database

Because functional food factors have been regarded as non-nutrient factors, they are not included in conventional food factor charts, and consequently there is a need for a database that enables estimation of intake in the same way as for nutrients. There have been plenty of studies carried out on the use of foods in disease prevention and health promotion, but most of these end up buried in journals and rarely reach the eyes of the general public. There is now an initiative underway to dig these studies out and gather them into a searchable database. Such functional foods need to be categorized according to evaluation criteria such as: (1) scientific evidence of benefits based on tests on humans; (2) scientific evidence of benefits based on animal experiments; (3) scientific evidence of benefits based on cultured cell experiments; (4) foods unable to be evaluated; and (5) no evaluation. Once such a database is created, it should be possible to estimate intake of functional food factors, and accordingly to formulate dietary guidelines and manage nutrient intake in a way designed to maintain health. It should also be possible for this information to be applied to design the molecular structure of functional food factors and to develop vegetables and fruits containing appropriate amounts of such factors.

2) Nutrigenomics

It seems likely that functional food factors work to prevent illness and promote health by serving as a kind of living signal to either directly or indirectly influence genes (particularly at transcription level). Nutrigenomics is the name given to the study of the effects of functional food factors on gene expression and regulation. Human intervention trials provide the most reliable data, but ethical issues and the recruitment of willing subjects make such trials difficult to implement. There are, moreover, a great many genes related to metabolism on the conditions, such as obesity and inflammation, hence, it is not easy to elucidate how each and every gene works. Nutrigenomics, however, uses DNA micro-array technology (DNA chips) that enables the study of the effects of food factors on the expression of hundreds and even thousands of genes in a single experiment. This in turn enables us to learn about the metabolism of functional food factors and to draw conclusions regarding where and how the resulting metabolic products work their effects.

Carrot retinoids and salmon unsaturated fatty acids are known to combine with specific genes through peroxisome proliferator-activated receptors (PPARs) to suppress the synthesis of fatty acids and promote fatty acid oxidation. Results of DNA micro-array experiments on rats fed a diet containing ω 6 and ω 3 fatty acids showed that these fatty acids affected the expression of over 300 out of 12,000 genes (Berger et al., 2002). Experiments to screen the effects of about 1,000 different plant extracts on cultured human cells have also shown that turmeric curcumin, grape resveratrol, green tea catechin, black tea theaflavin, vitamin E, and other compounds contain factors that help to suppress the expression of the COX-2 gene that is associated with inflammation (Subbaramaiah et al., 2000).

These are all discoveries revealed for the first time by nutrigenomics, and they could not have been obtained without the use of DNA sequencing technology and genomic data. Nutrigenomics also holds out promise for the discovery of either positive or negative effects of combining food factors, which are heretofore unanticipated functions in single food factors.

3) Personalized foods and nutrition guidance

Another area of application of gene technology is personalized foods and nutrition guidance. This application makes use of single nucleotide polymorphisms (SNPs) in the human genome. SNPs are DNA sequence variations that occur with a frequency of one in every several hundred nucleotide bases, with every individual possessing a different mix of SNPs. It has been demonstrated that many SNPs can have an influence on a person's receptivity to a certain drug, or predisposition to drug side-effects and so forth. Food factors have much weaker effects on physiology than medicines, making the elucidation of their effects difficult, but if individual differences in the disease prevention effect of foods or food factors with known functions can be elucidated, it should be possible to develop personalized foods and nutritional directions.

At Kagawa Nutrition University, nutrition clinic patients are tested for SNPs known to be involved in obesity and circulatory diseases and are investigated for any correlation between inherited traits and nutritional guidance or health indicators. Results showed that among the patients following nutrition clinic dietary guidance, there was no significant difference in body mass index (an indicator of degree of obesity: body weight^2/height) and total cholesterol between those possessing certain SNPs and those lacking them, proving that nutritional and exercise guidance is effective even for those possessing certain SNPs. It has also been shown through analysis of correlations between genetic constitution, diet, and health of various Asian peoples including Japanese that a far higher proportion of Japanese and Palauans than Caucasians possess SNPs of obesity-related genes such as leptin receptor and PPAR$_Y$ 2. This suggests that Japanese and Palauans have the kind of genetic make-up that enables them to weather a certain degree of nutritional deficiency better than Caucasians, and may, on the contrary, be more liable to put on weight if their nutritional intake is excessive.

Human health and functional foods

It is likely that further research will cast light on the influence of an increasing number of food factors on disease prevention and the maintenance and promotion of health, with the consequence that an increasing array of foods will be developed and put on the market. As long as the efficacy of functional foods is corroborated by solid scientific research and evidence, the suspicion surrounding present-day so-called health foods is likely to fade, and foods reflecting the new concepts will gain favor. In fact, as functional food science advances, health promotion and disease prevention functions (including lifestyle-related diseases) of an increasingly wide range of both new and traditional foods are likely to elucidated, and we may well witness the dawning of a new evidence-based age of convergence between food and healthcare.

But at the same time, the fact is that food is only food. I think we need to stop and ponder the meaning of the line in the preamble to the WHO's constitution that reads "Health is a state of complete physical, mental and social wellbeing, and not merely the absence of disease or infirmity". It is important to remember that the enjoyment of health is a fundamental right of every human being and that we must consider the health also of the elderly, the sick, infants, and otherwise physically weak people. It goes without saying that any foods that may bring a lot of benefits to physical health, but are detrimental to mental health are not acceptable. No matter how effective a food is in promoting health and preventing disease, if it is unpalatable or otherwise lacks appeal as a food, it is not likely to be around for long. This is because food performs more than primary, secondary and tertiary functions, possessing as it does the power also to provide emotional fulfillment and solace, foster communication with others, sustain culture, bind local communities, and in fundamental ways enable us to be who and what we are.

Part II
Alternative Medicine and Alternative Agriculture

8	**A Message from the Symposium Organizer** ·························· 39 *Tadayoshi Shiba*	
9	**Partnership between Alternative Medicine and Alternative Agriculture** ······· 41 *Katsu Minami*	
10	**Alternative Medicine: The Gap between Goal and Cause** ··················· 45 *Norio Yamaguchi*	
11	**Alternative Agriculture: Origin and Aims** ··························· 48 *Kazutake Kyuma*	
12	**Alternative Medicine and Oriental Medicine: In Search of Evidence by Scientific Elucidation** ························· 54 *Haruki Yamada*	
13	**Environmental Conservation Agriculture** ···························· 56 *Kikuo Kumazawa*	
14	**From the Production of Conservation Livestock Products to the Hospital Ward** ·· 62 *Tomiharu Manda*	

Chapter 8

A Message from the Symposium Organizer

Tadayoshi Shiba

I would like to start with some remarks on behalf of the organizer of Kitasato University's Second Agromedicine Symposium.

When people meet for the first time at international conferences and academic conventions, they always ask, "What's your field?" This seems to be the usual greeting for modern intellectuals. Especially in the United States and the countries influenced by its civilization, there are many situations in which people find it hard to become acquainted without this question. So it is that international conferences and other such events are held as an attempt to bring the varied expertise of many specialists to bear on problems.

Naturally, there are limits to each individual's abilities and lifetime. And that is why we focus on a particular subject area, belong to universities and research institutes, hold academic conferences, meetings, and other forums, become specialists, and publish our research and other achievements for posterity. That is also the 20th-century story of agriculture and medicine, which both belong to the life sciences.

For example, I chose biology. My work is analyzing, interpreting, and explaining living organisms. But in biology or any other field of enquiry, what we study does not exist in isolation. Each object of study has many facets and is connected to many other objects of study.

Take global warming for instance. Some of its aspects involve meteorological phenomena in the troposphere and stratosphere; other aspects include the metabolism of organisms using processes such as photosynthesis and also the combustion of biomass. In other words, global warming isn't the result of meteorological factors alone. It was the many environmental problems we experienced in the 20th century that gave many researchers a deep understanding of this.

Meanwhile, these experiences have given us a keen awareness of the need for comprehensive knowledge. Just as the Japanese word for "farmer" literally means "one who has knowledge about everything," farmers need comprehensive knowledge about things including the physiology of crop growth, soil science, meteorology, fertilizers, and topology. Biology originally started as taxonomy and then took in peripheral areas related to organisms overall, such as physiology, ecology, and molecular genetics.

But now such integrated specialists have disappeared, and specialists who might be called analysts continue to emerge in large numbers. Even though they have learned biology, if we take for example young researchers who have learned about biotechnology with a special microbe, usually they cannot fully answer contemporary complex questions about living organisms. But provide answers we must. And for that purpose, in the process of learning the basics as analysts, we must master the insights needed as integrated specialists.

I imagine this is happening also in medicine and agriculture, which both concern themselves with human life. Fortunately, Kitasato University's Second Agromedicine Symposium considers problems which modern medicine and agriculture cannot completely deal with and takes a fresh look at them from the perspectives of traditional medicine and alternative medicine or of alternative agriculture. In this way we are attempting to once again deepen mutual understanding and to find the path to collaboration between medicine and agriculture in our efforts at integrating our work regarding human life.

I hope that this symposium will offer meaningful and practical discussion and generate new thinking and suggestions about health issues through food and the environment. I would like to thank Kanazawa Medical University, the Cabinet Office, the Ministry of Education, Culture, Sports, Science and Technology, the Ministry of Health, Labor and Welfare, and the Ministry of Agriculture, Forestry and Fisheries for their assistance in holding this symposium, and I express my gratitude to all those who accepted with alacrity my requests to give talks.

Chapter 9

Partnership between Alternative Medicine and Alternative Agriculture

Katsu Minami

Introduction

Both medicine and agriculture, which concern themselves with life, have their "alternative" versions. Just as people working in healthcare are unfamiliar with the term alternative agriculture, those involved in agriculture find the term alternative medicine to be novel.

Are agriculture and medicine similar because they arose from the same roots, and even now walk similar paths? Looking back on ancient times, we see that in medicine there arose the "concept of sanitation" requiring people's cooperation, while in agriculture there arose "ceremonies" to pray for bountiful harvests and to avoid natural disasters, which made cooperation essential.

From the mid–19th century and into the 20th century, advances were realized in laboratory medicine, and attention was focused on the causes and prevention of epidemics. Biochemistry merged with molecular biology and became a powerful weapon in understanding life processes. In agriculture the manufacture of chemical fertilizers and pesticides began, and agricultural production grew by leaps and bounds. Molecular biology became a major field of endeavor, and genetically modified food plants were created.

In recent years modern medicine, which is based primarily on Western medicine, has seen the emergence of alternative medicine, which substitutes for and complements modern medicine, while intensive agricultural production, which uses mainly chemical fertilizers and pesticides, is similarly substituted for and complemented by alternative agriculture. These alternatives have characteristics of the life sciences. In the early 21st century, medical science has concluded the work of decoding the base sequences of the human genome, while agriculture has done likewise with the rice plant genome.

These days some people say we should consider medicine to be "social capital" that humanity holds in common, like education and the environment. This way of thinking is similar to the "one–third spring water" idea which conceived valuable spring water as common capital for agricultural production, and upon which the water was shared impartially. I would like to add that Hirofumi Uzawa defines "common social capital" in this way: "[Common social capital] means social measures which enable all the people living in a country or in a certain region to lead fulfilling economic lives, to develop superb cultures, and to sustainably and stably maintain a people-friendly society."

Is a partnership between alternative medicine and alternative agriculture possible?

Agriculture and medicine, which have a shared historical background as discussed above, face the challenges set forth by 21st century preventive medicine which include the assessment, management, and communication of risk, the prevention of illness, and improved quality of health. How can alternative medicine and alternative agriculture partner in response to these challenges in the field of medicine? Tackling this contemporary problem is of the greatest importance in responding to the needs of society.

A number of the achievements of 20th-century technological knowledge suggest that science and education under an agriculture?medicine partnership will be indispensable to our 21st century world. Representative of such areas are prevention of illness, health improvement, food safety, conservation agriculture, and healing agriculture. Despite the saying that medicine and food serve the same purposes, there has not been much emphasis on education and science under an agriculture-medicine partnership.

Until now there have been no opportunities to discuss how alternative medicine and alternative agriculture are related to these areas, so I hope we shall be able to do so at this symposium.

Recent trends in alternative medicine

To start with alternative medicine in Japan, the Japanese Society for Complementary and Alternative Medicine was established in 1998. Organizations in the United States include The National Center for Complementary and Alternative Medicine (NCCAM) and the Complementary Alternative Medical Association (CAMA). Books on the subject include *An Exhortation for Alternative Medicine* by Kazuhiko Atsumi and Teruo Hirose, The Alternative Medicine Handbook : *The Complete Reference Guide to Alternative and Complementary Therapies*, written by Barrie R. Cassileth and translated into Japanese by Kimiko Asada and Atsushi Hasegawa, and *Complementary and Alternative Approaches to Biomedicine*, edited by Edwin Cooper and Norio Yamaguchi.

Complementary and alternative medicine (CAM) was originally meant to complement modern medicine, which is mainly Western medicine. And in some countries CAM is used in the same meaning as traditional medicine. However, Chinese herbal medicine, Japanese/Chinese medicine, acupuncture, moxibustion, and the like, which in the West are classified as CAM, have long existed in Japan, China, and Korea, so in Japan they are not "alternative," but rather traditional medicine that is officially part of modern medicine. In response to the increase in chronic afflictions and lifestyle-related diseases in the developed countries, there is an awareness of the importance of not only cures, but also preventive measures, and there is increased demand for CAM. As a concrete way of answering this demand, the US enacted the Dietary Supplement Health and Education Act in 1994, which opened the way for actively exploiting the effectiveness of herbs.

But compared with modern Western medicine, the CAM field is still in a chaotic state lacking sufficient scientific verification. For that reason, a recent appearance is the concept of eCAM, or evidence-based complementary and alternative medicine, which aims to bring evidence-based order into this chaotic field.

There is an international journal on this field published by Oxford Journals. It has a note which reads: "Evidence-based Complementary and Alternative Medicine (eCAM) is an international, peer-reviewed journal that seeks to understand the sources and to encourage rigorous research in this new, yet ancient world of complementary and alternative medicine."

One of the authors of *An Exhortation for Alternative Medicine*, Teruo Hirose, translates "alternative medicine" into Japanese as daitai iryou in his papers.

Recent trends in alternative agriculture

Japan has a variety of agricultural methods that can be called alternative. One book that explicitly uses alternative agriculture in its title is the US National Research Council report *Alternative Agriculture* written by the Committee on the Role of Alternative Farming Methods in Modern Production Agriculture and translated into Japanese as *Alternative Agriculture : In Search of Permanently Sustainable Agriculture* (translation supervised by Kazutake Kyuma, Ryohei Yoshida, and Kazuo Nishimura). In the United States there are the American Society of Alternative Agriculture and the *American Journal of Alternative Agriculture*.

Throughout the world, individuals and organizations are researching, and endeavoring to develop and deploy, a system that will arrive at a single type of agriculture that will increase soil productivity, conserve the natural environment, efficiently use land and resources, and lower production costs.

Behind this aim is the stern reality that agriculture has a negative impact on the environment and that that impact is of vital importance to policymakers, farmers, and consumers. Problems are pollution of the soil, groundwater, and air by pesticides, chemical fertilizers, and livestock waste. There is also the matter of safety with respect to pesticide residue and buildup in crops and food. Other problems actually occurring include soil erosion, soil salinization, and the depletion of ground water sources used for irrigation.

In 1989, the US National Research Council report *Alternative Agriculture* was published in response to these problems.

Alternative agriculture does not signify one certain way of farming, but rather includes a variety of systems ranging from organic systems which use no synthetic chemicals at all to those which cautiously use pesticides and antibiotics to control certain diseases and insects. As such, alternative agriculture is agriculture preceded by a modifier such as biological, low-input, organic, regenerative, or sustainable.

Some examples are integrated pest management, low-intensity livestock production, crop rotation systems, and tillage methods that decrease soil erosion. Therefore, alternative agriculture is that which seeks to incorporate these technologies into the way agriculture is performed.

One kind of alternative agriculture is conservation agriculture. The Sustainable Agriculture Office of the Ministry of Agriculture, Forestry and Fisheries defines it as "sustainable agriculture which gives consideration to mitigating environmental burdens, such as the use of chemical fertilizers and pesticides, while taking advantage of agriculture's material cycle function and paying attention to matters including harmonization with productivity." Generally, this means agriculture and agricultural methods which burden the environment as little as possible.

The Japanese word for alternative agriculture (*daitai nogyo*) was first used by Kazutake Kyuma and the other translators of the National Research Council report *Alternative Agriculture*.

The purpose of this symposium is to consider problems which modern medicine and agriculture cannot completely deal with and to take a fresh look at them from the perspectives of traditional and alternative medicine and of alternative agriculture. In this way we attempt to once again deepen mutual understanding between medicine and agriculture and examine the specific challenges of each, and then by this means find the path to collaboration. As one example of the agriculture-medicine partnership, we will describe the flow of conservation livestock products from production to hospital ward, which we are already doing at Kitasato University, and use this as one reference point for considering how alternative agriculture can respond to medicine.

Chapter 10

Alternative Medicine : The Gap between Goal and Cause

Norio Yamaguchi

Many of the incurable diseases left to the 21st century, such as cancer, AIDS, allergies, and autoimmune diseases, are not completely curable with the one-dimensional approach of Western medicine, and that is why people are looking outside of Western medicine for medical care to alleviate symptoms and ways to mitigate subjective symptoms. As a whole these initiatives are called alternative medicine. Definitions in the West and Japan refer to healthcare outside of the medical science and medical treatment that is taught in medical educational institutions. Into this category go all types of regional medicine, of which Eastern medicine tops the list. Except for health supplements, bathing in hot springs, aroma therapy, and other cryptic and scientifically unverified treatments, there are 43 categories listed in the Burton Goldberg Group's *Bible of Alternative Medicine*.

The expressions "alternative medicine" and "complementary medicine" are terms that evaluate "Eastern medicine" and the like in the West. Here too one can discern a stance which puts Anglo-Saxon thinking first, where modern medicine that started in the West is primary, and everything else is secondary. Alternative medicine is getting another look especially in the United States because of factors which might be characterized as second thoughts engendered by the dead end in modern medicine, the reality of treatment limitations, and doubts about the structuring of Western medicine, which is based on modern Western science to date. I think another underlying reason is the accelerating return-to-nature trend as a reaction to the overdevelopment of modern civilization. American alternative medicine is a mixture of wheat and chaff. It includes some incomprehensible elements and should not be uncritically accepted. It seems there is also resistance to putting these on the same level with Sino-Japanese medicinal science, which is Japan's medical tradition.

Further, traditional medicine does not merely offer treatment modalities that complement modern medicine, because the approach of traditional medicine is based on substance and ideas that will reform the approach of modern medicine. In other words, there are independent medical care systems that exist in the lands of various cultures, and the supplementing of each by the others is what makes global medical theory and practice possible.

Complementary and alternative medicine will henceforth have a broad scope and include not only traditional medicine and folk remedies from around the world, but also new treatment methods not covered by health insurance. Specifically these are Chinese medicine (Chinese herbal medicine, acupuncture, moxibustion, shiatsu, qigong), Indian medicine, Middle Eastern medicine, immune therapy, medicinal foods and health supplements (anti-allergy foods, immune-stimulating foods, preventive and supplementary foods, etc.), herbal treatment, aroma therapy, vitamin treatment, trace elements, dietary therapy, psychiatric and psychological therapy, hot spring therapy, oxygen therapy, and others. Certainly, these include treatments that are unscientific and are hard to accept by physicians who practice Western medicine, but it is also a fact that the number of treatments whose action mechanisms and effectiveness have been scientifically demonstrated is rapidly increasing.

At this time, Western countries are taking the lead in activities to properly evaluate alternative medicine. Western countries give Eastern medicine as their number one option for alternative medicine. It was against this backdrop that the Asian-led international medical magazine *eCAM* was launched primarily by Japanese researchers. The purpose of *eCAM* is to look widely for treatments, medicines, and traditional food — medicine preparations, and use the same judgment criteria as in Western medicine to assess, select, and inform the public about them.

The practitioners of Eastern medicine, Indian medicine, Middle Eastern medicine, and other traditions around the world advocate the advantages and legitimacy of their own countries' medical traditions from their own stances. If these could all be compared with a common yardstick, it would be possible to develop a hybrid medical science in which each takes advantage of the others' characteristics.

When the new international journal *eCAM* was launched, it proposed immunological elements as a common yardstick by which to evaluate alternative medicine, and worldwide agreement was obtained. Specifically, the evaluation criteria are peripheral leukocyte, granulocyte subgroups, and the quantitative and qualitative attributes of lymphocytes. In our laboratory we have used immunological elements to prove the efficacy of hot spring bathing, acupuncture and moxibustion treatment, music therapy, exercise therapy, and aroma therapy. About one of these, hot spring bathing, we have released results showing that while long-term bathing is of course good, even two-day short-term use is quite effective.

If we look for the origins of Eastern medicine, we find that Japan, China, Korea, and Taiwan all qualify, but none serves as a temporally uninterrupted source because of gaps in the presence of each in the Eastern medicine timeline. In response to the rising worldwide interest in alternative medicine, the World Health Organization wants to choose a country to administer the formulation of a model method for the technical level of Eastern medicine, especially acupuncture and moxibustion. East Asian countries to which this is applicable have reacted to this with sensitivity, and each is making claims for its own legitimacy and advanced status.

Japanese medicine, which quickly incorporated Western medicine in the Meiji period (1867-1912), understands its advantages and shortcomings and is the most advanced in East Asia. For that reason it makes sense for Japan to propose itself as the country to administer

Eastern medicine.

There was a time when Eastern medicine, especially Chinese herbal medicine, was the mainstream of medical science in Japan. Because the Meiji government totally converted the medical care system to Western medicine as a matter of national policy, Eastern medicine barely remained in existence until the end of World War II. On the other hand, starting in Japan's Edo period (1600-1868), Western medicine was gradually introduced into Japan while undergoing a transition from Dutch to German medicine. Now, after the passage of about a century, North American medicine has become the mainstream in Japan. Thanks to this history, medical care in Japan has been able to benefit from Western medical science, and over the last 100 years our average life span has lengthened considerably. We find that Japan's basic medical science has many fields that are on the cutting edge of progress.

Japanese medicine, which quickly incorporated Western medicine in the Meiji period (1867-1912), understands its advantages and shortcomings and is the most advanced in East Asia. Take acupuncture, for instance. In East Asia this is the least invasive, and also uses needles only once. Therefore it has the least danger of invasive infection. The acupuncture and moxibustion in Western countries was taken there from various countries by individuals. In consideration of that situation, it makes sense that Japan, a candidate for the Asian administering country, should take the lead.

However, if you search Medline for the number of articles about clinical research on acupuncture and moxibustion, those from Western countries account for the overwhelming majority, while the number from potential Asian administering countries is incomparably small. This is not because of a lack of effort to publish articles, but because articles are not in a format for publication in international journals. If this state of affairs remains unchanged, Asian countries will be heavily influenced by the treatment policies set forth in Western journals, even though they are technically crude. If this becomes the trend, it is only a matter of time until the same thing starts happening to diagnosis, treatment, and research in Japanese and Chinese herbal medicine.

This does not mean that Western countries are deliberately trying to be in the driver's seat with respect to Eastern medicine. Sometimes they recognize Japan's qualification to take the lead. Publication of the international journal *eCAM* has come about precisely because academics from around the world have approved of Japanese leadership in the CAM field. People are hoping that the world standard will be alternative medicine that was founded in the East and refined there by the sensibility gained from Western medicine, not the Eastern medicine of short experience that has taken root in the West. It is necessary that through seminars such as this one we make a strong appeal to nearby countries and the rest of the world for Japan to establish itself as the administering country of Eastern medical science.

Chapter 11

Alternative Agriculture : Origin and Aims

Kazutake Kyuma

Development of agriculture prior to the origin of alternative agriculture

In the age when humans obtained their food by hunting and gathering, their environmental impact was negligible. Beginning roughly 10,000 years ago when agriculture arose in various places around the world, humans have seriously disrupted ecosystems, opened forests and grasslands for farmland, and cultivated grains and many other crops. Further, not only have they grazed livestock on natural grasslands, they converted large tracts of land into pasture and raised large numbers of livestock. Irrigated agriculture, which draws water into arid lands, is almost as old as agriculture itself and supported the ancient civilizations of Mesopotamia and the Indus Valley.

Without a doubt, agriculture began as a destroyer and modifier of nature, and throughout its history has scarred the earth around the world. It is known that since ancient times, many civilizations have declined owing to the collapse of the food production bases that supported them. Nevertheless, it is also a fact that since the beginning of agriculture people have learned from many experiences such as this and have created highly sustainable agricultural ecosystems that could be called an alternate nature. European mixed farming, which combines crops and livestock, came to have a high degree of sustainability on a foundation of crop rotation based on the characteristics of each locale, while the monsoon Asia agricultural system based on wet rice agriculture made it possible to support a high population density because of its high productivity and stability.

Against the backdrop of mining and manufacturing development brought about by the industrial revolution and advances in science and technology with the coming of the 19th century, many new off-farm materials were introduced into agriculture. Starting with the Chile saltpeter and guano brought from South America, next came mined potash ore and phosphate rock, which were later to be chemically processed into superphosphate of lime and other products used in agriculture, thereby helping to greatly raise the productivity of agricultural soil. Steam-powered farm machinery appeared at about the end of the 19th century, and by mechanizing agricultural tasks that until then had been performed exclusively by draft animals, they promised a quantum leap in labor productivity. As European and American agriculture thus entered the 20th century, they were growing out of pastoral, self-sufficient agriculture and preparing for the transition to modern intensive agriculture.

Origins of alternative agriculture

The success of industrial nitrogen fixing in the early 20th century led the subsequent conversion to chemical agriculture in the developed countries. Fixing atmospheric nitrogen and making it into chemical fertilizer released agriculture from dependence on livestock manure and compost. This combined with advances in mechanization to increase the productive capacity of farmland, and through changes such as lightening the labor burden on farming families, became the first step toward big changes in the state of agriculture. DDT, BHC, and other chemically synthesized insecticides started in the 1930s, and proved themselves very useful in preventing insect damage to crops which until then had depended on natural insecticides such as Dalmatian chrysanthemums. In time many kinds of synthesized pesticides were developed to deal with diseases and weeds, and these came to be widely used.

The mechanization and use of chemical inputs in agriculture promoted the enlargement of scale and increased efficiency of agriculture in Western industrialized countries in the second half of the 20th century, which in turn led to monocultures and successive cropping of commodity crops, ultimately achieving unprecedentedly high yields such as 10 t/ha of corn and 8 t/ha of wheat. These monocultured grains came to be overproduced, and during the Cold War years were very influential as strategic goods in international politics and diplomacy.

The trend toward ultra-intensive agriculture, which is characterized by the mechanization, use of chemical inputs, monocultures, and successive cropping in Western countries, started in the 19th century and became pronounced in the second half of the 20th century, but came to a dead end in the 1980s when overproduction engendered a market slump in agriculture. At the same time, a number of internal problems that had built up in agriculture became apparent. In particular, environmental pollution, such as pollution of groundwater, by chemical inputs, and also doubts about the safety of food because of those same inputs, generated public distrust and anxiety toward agriculture. The 1985 Food Security Act (Farm Bill) of the United States contained provisions for research on agricultural productivity and advocated raising environmental awareness among farmers. Agricultural reforms in the EC (now the EU) that same year introduced measures including those which encourage deliberately extensive farming and letting land lie fallow.

Although alternative agriculture arose as a response to these social and policy trends in the developed countries, especially the United States, the seeds of alternative agriculture were sown in the first half of the 20th century by the practitioners of organic agriculture who did not feel right with, for example, the "modernization" of agriculture, so such practices are not necessarily new. However, because the term alternative agriculture was used in counter-position to the conventional agriculture of the day, there is no doubt that the reference standard used was the agriculture generally practiced in Western countries in the 1980s. Thus alternative agriculture can be seen as embracing a broad range of agricultural methods, from orthodox organic agriculture to deliberately limiting the use of chemical fertilizer and synthesized pesticides in consideration of the environment or with the aim of reducing costs through low inputs. But considering the fact that alternative agriculture is often used in the same sense as sustainable agriculture, it is perhaps all right to limit the term alternative agriculture to something with a strong awareness of agricultural and social sustainability through consideration for the environment.

The term alternative agriculture itself started with the Henry A. Wallace Institute for Alternative Agriculture, which was founded in 1983. In 1986, the institute started publishing the *American Journal of Alternative Agriculture*. About the same time, the National Research Council launched a special committee that carried out a broad study on the state of alternative agricultural methods and on the policy, business management, and technical factors related to their dissemination. The results were reported in a 1989 report entitled Alternative Agriculture (National Academic Press). Its Japanese translation was published in 1992 as *Alternative Agriculture : In Search of Permanently Sustainable Agriculture* (translation supervised by Kyuma, Kada, and Nishimura; published by the International Nature Farming Research Center). This is likely the first time the term alternative agriculture was used in Japan.

Aims of alternative agriculture

According to the report of the National Research Council, alternative agriculture refers to any agricultural production system that systematically pursues the following goals :
- More thorough incorporation of natural processes such as nutrient cycles, nitrogen fixation, and pest−predator relationships into the agricultural production process;
- Recduction in the use of off−farm inputs with the greatest potential to harm the environment or the health of farmers and consumers;
- Greater productive use of the biological and genetic potential of plant and animal species;
- Improvement of the match between cropping patterns and the productive potential and physical limitations of agricultural lands to ensure long−term sustainability of current production levels; and
- Profitable and efficient production with emphasis on improved farm management and conservation of soil, water, energy, and biological resources.

It is safe to say that advocating practices such as taking into account the potential productive capacity and natural characteristics of farmland, diminishing to the maximum extent the use of off−farm inputs, and making the maximum use of natural processes such as nitrogen fixing and natural enemies is in the starkest contrast with the ultra−intensive farmland management methods used in conventional agriculture. By these means, the ultimate aim of alternative agriculture is to benefit farmers economically and to benefit the nation with qualitative improvement to the environment.

Below are some of the specific techniques and methods used in low−input alternative agriculture:
- Crop rotation that mitigate weed, disease, insect, and other pest problems; increase

available soil nitrogen and reduce the need for purchased fertilizers; and, in conjunction with conservation tillage practices, reduce soil erosion.
- Integrated pest management (IPM), which reduces the need for pesticides by crop rotations, scouting, weather monitoring, use of resistant cultivars, timing of planting, and biological pest controls.
- Management systems to control weeds and improve plant health and the abilities of crops to resist insect pests and diseases.
- Soil- and water-conserving tillage.
- Animal production systems that emphasize disease prevention through health maintenance, thereby reducing the need for antibiotics.
- Genetic improvement of crops to resist insect pests and diseases and to use nutrients more effectively.

As one can see from these examples, alternative agriculture uses no techniques or methods which could in particular be called new. In fact, as shown by crop rotation, one salient aspect of alternative agriculture is the return to basic approaches and techniques which were originally part of traditional cropping and animal husbandry practices.

Evaluating alternative agriculture

The National Research Council sums up the results of its broad study and investigation of alternative agriculture methods as follows. When management is done well technically, the income and profitability of farmers practicing alternative agriculture are not inferior to those of farmers practicing conventional agriculture, although the report says that more information, labor, time, and management expertise are required. The report is critical of government policy, saying that often subsidization policies are an impediment to the adoption of alternative agriculture. Concerning research and education, the report provides a dose of bitter medicine by saying that the interdisciplinary nature and effectiveness of research and education has been lost due to specialization of agricultural research, and that the environmental and social costs of agriculture cannot be accurately assessed.

The Potash and Phosphate Institute, which manufactures and sells fertilizer, is of course critical of alternative agriculture. It expresses doubts about its sustainability by claiming for example that increasing the frequency of tillage to reduce herbicide use could increase topsoil loss due to erosion and worsen the pollution by surface runoff, and that reducing phosphate and potash application will lower the nitrogen-fixing ability of pulses used between crop rotations, thereby leading to reduced yields.

Some researchers are likewise critical of alternative agriculture. Loomis and Connor (*Crop Ecology : Productivity and Management in Agricultural Systems*, Cambridge University Press, 1992), for example, say that low-input agriculture has low solar energy conversion efficiency, making it impossible to efficiently use scarce nutrient and water resources; alternative agriculture ignores the fact that intensive agriculture has lower energy consumption per unit production than does extensive agriculture; and that crops produced with low-input agricultural methods are sold to consumers at a premium, but that the higher price is like a subsidy paid to low-input farmers by society at large as farmland management fees. They claim that to ensure that agriculture will produce enough food for humanity, it is essential to have off-farm inputs such as fertilizer and pesticides, as well as intensive modern agriculture that makes use of advanced technologies. These advanced technologies include, according to the authors, quantitatively and qualitatively enhancing agricultural information, bettering the understanding of energy, soil resources, genetic resources, weeds, diseases, and insects, and improving management. Regarding soil management, for instance, their view is that accurately determining where and when to apply fertilizer makes precise fertility management possible through the use of chemical fertilizers, which leads to modern precision agriculture.

The result of a study performed in the first half of the 1990s in Iowa showed the reactions of farmers in general to alternative agriculture. While their reasons for not adopting alternative agriculture included the conflict with government subsidy policy, they were predictably concerned about "declining yields," "worsening weed damage," "lower profitability," and other problems, and often cited "the need for more knowledge, information, technology, and the like for management." This is none other than the antithesis of how conventional large-scale monocropping agriculture became widely adopted: farmers would be guaranteed high yields and high profits as long as they just planted a single crop, used machinery specialized for that crop, used manufactured fertilizers, and followed a specified pest management regime. This concern about alternative agriculture shows that, to practitioners of conventional farming, the need to acquire a large amount of knowledge, information, and technology is a high barrier to the adoption of alternative agricultural methods.

The potential of alternative agriculture

As there are various ways to be heedful of the environment, it is hard to assess exactly the extent to which alternative agriculture has been adopted. However, in the United States the low-input sustainable agriculture program (LISA) launched by the US Department of Agriculture in 1988 was later changed to the sustainable agriculture research and education (SARE) program and continued to be heavily promoted. Not only that, the government intensively backs alternative agriculture, such as by eliminating policy barriers that were criticized for hampering the adoption of alternative farming methods. Additionally, there are many spontaneous movements such as for organic agriculture in the private sector, and it seems that the base for alternative agriculture is definitely expanding. There are also movements which involve not only producers but also consumers: Produce subscriptions between small family farms and consumer groups, which are said to have started with Japan's producer — consumer partnering movement, are expanding as community-supported agriculture (CSA), and consumers are showing a preference for locally grown produce in accordance with the idea of local production and local consumption. These developments show that social conditions surrounding alternative agriculture are clearly different from those in the 1980s when the movement started.

In the EU, the 1985 farming reforms never went beyond dealing with the superabundance of agricultural produce, but the 1992 Common Agricultural Policy slashed production subsidies for certain crops, introduced direct income guarantee (decoupling), and worked out environmental conservation measures for rural communities in especially disadvantageous geographical areas. In 2003, the EU went further in this direction and developed policy which increases direct payments that contribute to areas such as emphasizing environmental conservation, food safety, and livestock welfare. Because Europe traditionally has a strong predilection for organic agriculture known variously as "ecological agriculture" or "biological agriculture," it seems that the private sector too is practicing a variety of alternative farming methods under the policy inducements described above.

Twenty years since its inception, alternative agriculture is making steady progress.

Chapter 12

Alternative Medicine and Oriental Medicine: In Search of Evidence by Scientific Elucidation

Haruki Yamada

Complementary and alternative medicine is a term which arose from the Western perspective to signify medical treatment which contrasts with treatment based on Western medicine, which is now the world mainstream, and which complements this mainstream medicine or takes a different approach. Alternative medicine as defined in the West includes various treatments such as traditional medical treatments, chiropractic treatment, psychotherapy, herb therapy such as aroma therapy, and treatment with health foods. Also included are the traditional herbal treatments, acupuncture and moxibustion, and other treatments in Oriental medicine, which have been used since ancient times as traditional medicine in Japan, China, Korea, India, and other Asian countries. Japanese traditional medicine includes Kampo medicine, acupuncture, and moxibustion. Especially different from China, Korea, and other countries is that these treatments are integrated with Western medicine, and physicians use them under legal protection, as with Western medicine. Thus, in Japanese medical treatment it is permissible for a physician to practice both Oriental and Western medicines. In that sense, Kampo medicine, acupuncture, moxibustion as the traditional medicines are not termed alternative medicine from the Japanese point of view, but in the Western point of view these are seen as approaches to treatment found in alternative medicine. But in the final analysis Western countries are arguably moving in the direction of the integrated medical treatment already used in Japan.

Alternative medicine gives people more choices for treatments. Japan, which has been using traditional medicine since ancient times, must play an international leading role in which it can constantly provide information on alternative medicine. Just as with alternative medicine in the West, globalization makes it necessary to proceed further with scientifically demonstrating the clinical efficacy of Kampo medicine, acupuncture, and moxibustion, and build a body of evidence. At the same time, scientific demonstration research for building evidence of Kampo medicine's effectiveness can also be used as the methodology of research for building evidence of the other herbal treatments and treatments in alternative medicine. Kampo medicine is a multi-component system in which multiple herbs are prescribed, and the interactions of the herbs yield diverse pharmacological effects differing from those of new drugs. These complex-system drugs regulate the complex biological systems which involve the immune system, nervous system, endocrine system, and others, making it necessary to elucidate the mechanism of action on the body's complex systems. Research on Kampo medicine includes research on Kampo medical science and Kampo pharmaceutical science. As part of the progress in science and technology, the total human genome has been decoded, leading us into the post-genome era, and cutting-edge medicine is advancing to medical care that is tailored to each individual patient. Kampo medicine has always emphasized treatment for the individual, and the Kampo medicines used for treatment are chosen after determining the patient's physical constitution and total physical state. In that sense as well, advanced medical care is on the way to becoming like Kampo medicine.

We need clinical research to translate into the language of modern medicine from the unique Oriental medicine methods used in Kampo medicine to determine a patient's condition. Research on Kampo pharmaceutical science needs to scientifically shed light on the mechanism behind Kampo medicine's clinical efficacy and its active ingredients. Research methods must also be developed. Feedback of basic research results to clinical practice holds forth promise of use in new clinical applications. Such an example is the discovery of Kampo medicines which improve brain function in the case of Alzheimer-type cognitive impairment. Also discovered is Kampo medicine that is effective against influenza virus infection by virtue of regulating the immune system. It is crucial for reproducible medical treatment that the crude drugs derived from the plants and other materials used in Kampo medicine are of high quality, and that is impossible without partnership between those involved in the fields of agriculture and medicine.

Attention is focused on the expansion of the health food market as an area of interest for alternative medicine, but the body of evidence for the effects of health foods varies in quality. Many products feature exaggerated advertising. We must gather better evidence and provide consumers with correct information.

It is said that humans selected as medicines those food items which are effective against illness. The Tang Dynasty medical book Qianjinfang says, "The basis of health maintenance and illness prevention is food, while that which quickly treats illness is medicine." And the Zhouli, which is said to record the institutions of the Zhou Dynasty (11th through 8th centuries BC), divides physicians into four areas of specialty. Among them, the "food physician" provides dietary treatment and is ranked higher than actual physicians. This is the basis of the concept that food and medical treatment are the same, which emphasizes preventive medicine. Kampo medicine has the idea of "curing before illness," and now with the greying of society, there are high hopes that alternative medicine, which includes Oriental medicine and functional foods, will be deployed in preventive medicine. In the field of veterinary medicine, livestock, pets, and other animals are already being administered Kampo medicine, acupuncture, and moxibustion treatments as Oriental veterinary medicine. The partnership between agriculture and medicine has already begun in connection with food, animals, plant materials, and other areas.

Chapter 13

Environmental Conservation Agriculture

Kikuo Kumazawa

Environmental conservation agriculture and sustainable agriculture

Environmental Conservation agriculture in Japan is defined as "sustainable agriculture which, through soil improvement and other means, takes care to mitigate environmental impacts including those caused by the use of chemical fertilizers and synthetic pesticides, while maintaining and promoting the natural cyclical function of agriculture and being mindful of balance with productivity and other considerations" (MAFF, GOJ, 1992). Arguably, this specifies a kind of sustainable agriculture which can be practiced at a very low level of food self-sufficiency (40% on a calorie basis in 2006) even while being an industrialized country.

Sustainable agriculture will play a part in the sustainable development of human society.

"Sustainable development" is defined as "development that meets the needs of the present without compromising the ability of future generations to meet their own needs" (UN World Commission on Environment and Development, 1978). Sustainable agriculture has three requirements: (1) it must be economically viable, (2) it must conserve the environment, and (3) it must be socially acceptable.

Conservation agriculture and the "environment"

The "environment" in Environmental Conservation Agriculture does not mean just the agricultural environment, but also includes the human and natural environments. Accordingly, Environmental Conservation Agriculture must at the same time conserve such things as soil, water, biological resources, people's living environments and food environments, and biodiversity.

Conservation of soil, water, and biological resources is linked to not only conserving the basic conditions for agricultural production, but also to the maintenance and upgrading of the multifunctional roles of agriculture. These include maintenance of soil fertility (that is, the productivity of the soil) by returning agricultural and livestock waste to the soil; maintaining irrigation and drainage systems; maintaining groundwater quality ; preventing eutrophication of rivers, lakes, and other bodies of water; conserving biodiversity; and preserving village woodlands.

Conserving people's living environments and food environments includes the cyclical use of solid wastes, and supplying safe, worry-free agricultural produce.

Environmental impacts of agriculture

Agriculture arose in the Neolithic age about 10,000 years ago. It created a gradually increasing burden on the environment, such as the release of soil carbon into the atmosphere and the growing loss of topsoil due to the opening of forests and other primal lands. In time, farmers encountered a crisis in the decline of soil productivity, but the development of intensive agriculture thanks to the invention of chemical fertilizers and the use of pesticides in the early 20th century enabled farmers to maintain food production to match population growth, and the invention of synthetic pesticides in the second half of the 20th century has made a large contribution to stabilizing agricultural production.

On the other hand, the enlargement in scale, mechanization, and monocropping that was meant to raise economic efficiency brought additional problems. It built a divide between crop farming and livestock, which lessened the application of compost and manure to cropland, and increased dependence on chemical fertilizers to compensate. Damage from successive cropping, disease, and insects also increased, and dependence on chemically synthesized pesticides increased to deal with it.

The Fertilizer Control Law and the Pesticide Control Law take into account the direct and indirect impacts that fertilizers and pesticides have on people and other living things and legally control their effectiveness for agricultural production and regulate their production and use. Nevertheless, it is now observed that the overuse of chemical fertilizers and chemical pesticides is having various environmental impacts.

1) Environmental burden due to fertilizer

The components of fertilizer that are problematic with respect to burdening the environment are mainly nitrogen and phosphorus, especially nitrogen. The nitrogen in organic and inorganic nitrogen compounds applied as fertilizer to farmland is absorbed and used by crops, but nitrogen supplied in excess of what crops absorb stays in the soil, and some of it runs off and outside the system as nitrate nitrogen. It first enters shallow or deep groundwater, then spring water, rivers, and lakes.

Water is essential to humans for drinking and other domestic uses in everyday life. Especially for drinking water, there are standards which take impacts on human health into consideration. The standard for nitrogen is "10 mg/L of nitrate and nitrite nitrogen." The main basis for this standard is the appearance of methemoglobinemia, or blue baby syndrome in infants and small children, but for adults it is not clear what disorder nitrate or nitrite nitrogen (below, nitrate nitrogen) induces, and one must take note that no standards have been established for the content of nitric acid in vegetables or other foods.

In Japan the environmental quality standard for nitrate nitrogen in groundwater is the same as that for drinking water.

In Japan nitrate nitrogen in groundwater exceeds the standard in around 5% to 6% of the total wells tested during 1995–2004. One of the wells with nitrate nitrogen concentration exceeding the standard had a high value of 77 mg/L.

In Japan too, it was reported in 1996 that an infants who drank well water with high nitrate nitrogen concentration of 32.2 mg/L developed methemoglobinemia.

Although nitrate nitrogen contamination of groundwater is seen all over Japan, most of it is seen in places with dry fields and orchards, and in farming villages. It is comparatively less in areas with rice paddies, but it is heavy in places with much livestock. Nitrate nitrogen in groundwater presumably comes from heavy fertilizer application on vegetables, tea, and orchards, and from livestock wastes.

2) Environmental burden due to pesticides

Concern about the environmental burden due to pesticides has been high since Rachel Carson's 1962 book Silent Spring, and in Japan it is well known that events such as the publication of Sawako Ariyoshi's book Fukugou Osen ("Compound Pollution") triggered the groundswell in the organic agriculture movement and other changes. The 1996 book Our Stolen Future by Theo Colborn and others informed the public about the broad impacts of pesticides on humans and wild animals. There is also a great deal of concern about the use and control of pesticides because of new-found knowledge such as the existence of pesticides contaminated with dioxins, which are powerful toxins.

Pesticide impacts include 1) direct effects on pesticide users, 2) effects on the health of people who eat agricultural produce with residues of pesticides used on them, 3) direct impacts on microorganisms and plants and animals in general from pesticides released into the environment, and 4) long-term biological effects from biomagnification and the food chain, or impacts on biodiversity.

There are supposedly no problems with food safety or effects on human health as long as pesticides registered and permitted under the Pesticide Control Law are used in compliance with safe usage standards, but because there are still problems with the overall environmental burden, including the impacts on biodiversity, it is necessary to limit the use of pesticides as much as possible.

Environmental conservation agriculture and building a cyclical society

1) Effect of soil improvement

Considered especially important in Environmental Conservation Agriculture is soil improvement; that is, measures to maintain soil productivity. While both organic and inorganic materials are used in soil improvement, considered important here is "agriculture's material cycle function" or the "natural cyclical function." As such, one thinks first of using organic materials such as compost and manure, and organic fertilizers.

Organic substances applied to the soil evince their effectiveness in terms of the soil's physical, chemical, and biological properties. Humic acid, a stable compound which decomposes with difficulty, is useful for many reasons including maintaining the formation of the soil's aggregated structure, magnifying soil cation exchange capacity through formation of clay – humus complexes, and maintaining buffer action. Soil is home to many microorganisms and animals of all sizes. Their actions and interactions decompose and use organic substances, while intermediate products and microorganism secretions have a variety of functions such as plant growth promotion and mutual inhibition of microorganisms. Especially complex microbial compositions make for healthy soil and have actions such as suppressing the proliferation of disease – causing bacteria. Further, they manifest the strengths of many beneficial bacteria such as those which fix nitrogen.

Crops which grow in soil like this have excellent root growth and activity, as well as good above – ground growth and fruit maintenance, and are healthy crops with resistance to disease and insects.

In comparison with crops using chemical fertilizers, crops grown in soil improved with organic materials and using organic fertilizers are generally better in quality.

Therefore the basis of Environmental Conservation Agriculture is soil improvement with organic materials, and healthy soil thus made is the foundation of healthy, good – quality crops.

2) The biomass nippon strategy and environmental conservation agriculture

The mass – production, mass – consumption, mass – disposal system which is the concomitant of human society's development invites the exhaustion of petroleum and other fossil resources, as well as mineral resources. At the same time huge volumes of organic and inorganic wastes have caused a variety of environmental problems, including a shortage of final disposal sites, and have hampered the development of a sustainable society.

These huge volumes of organic wastes were originally released into the environment after the organic materials formed by plant photosynthesis using atmospheric carbon dioxide and inorganic nutrients in the soil had been used in the so – called food chain process that nourishes animals, microorganisms, and humans. By returning these organic wastes to the soil, they are restored to their original carbon dioxide and inorganic materials through final decomposition, thereby completing the natural cycle. This is, in other words, nature's environmental cleanup action.

In 2000, the Basic Law for Establishing a Recycling – based Society was enacted in Japan, under which efforts are made to promote reduction, reuse, and recycling of wastes.

Organic wastes account for a high percentage of total wastes. Most can be processed using nature's cycle or "agriculture's material cycle function," but the core of that effort is Environmental Conservation Agriculture, which improves soil by actively promoting the composting of organic wastes.

The "Biomass Nippon Strategy" created in 2002 considers all organic materials that are not petroleum or other fossil resources to be biomass. The Strategy provides for the use of biomass as an effective resource for processing within the natural cycle, including its use for energy. Local Environmental Conservation Agriculture is seen as the core of local resource cycling.

Advancement of environmental conservation agriculture

1) Environmental conservation agriculture initiatives

A gradually increasing number of farmers are tackling Environmental Conservation Agriculture. A 2002 survey found that they accounted for 16.8% of commercial farmers, but only 69.8% of those farmers were using compost for soil improvement.

At that point in time, farmers growing organic produce had a strong desire to expand production, but among other farmers practicing Environmental Conservation Agriculture, those who wanted to maintain their present level outnumbered those who wanted to expand their efforts. These results suggest that a number of barriers, especially the small economic advantage, still stand in the way of undertaking Environmental Conservation Agriculture.

On the matter of soil improvement, gathering rice straw and using it for animal feed, spreading manure on farmland, and other practices encouraged through the stronger coupling between crop and livestock farming are commendable as one component in expanding Environmental Conservation Agriculture.

2) Sustainable farming law and eco-farmers

The "Law for Promoting the Introduction of Sustainable Agricultural Practices" (Sustainable Farming Law), enacted in 1999, is meant to promote conservation agriculture, and it sets forth the following technical particulars for Environmental Conservation Agriculture:

(1) Soil improvement techniques: Application of compost and other organic materials, use of green manure crops.

(2) Techniques to reduce chemical fertilizer use: Localized application, fertilizer application that regulates efficiency, organic fertilizer application.

(3) Techniques to reduce use of chemically synthesized pesticides: Hot − water seed disinfection, mechanical weeding, using animals for weeding, using biological pesticides, using antagonistic plants, cultivating resistant cultivars and rootstock, thermal soil disinfection, use of light, cover crops, use of pheromones, and mulching.

For each of Japan's prefectures, the law sets forth specific technical guidelines for each crop. These guidelines are tailored to the conditions of each geographical growing area and include the techniques listed above.

Farmers who purposely adopt such agricultural technologies are called "Eco-farmers." As of March 31, 2006 there were 98,875 Eco-farmers.

Support for environmental conservation agriculture

Generally, Environmental Conservation Agriculture is not what is called value-added agriculture, so producers are saddled with labor, costs, and other burdens more than in conventional agriculture. As such, getting more farmers into Environmental Conservation Agriculture requires concrete help with management, not just mere moral support. Various ingenious measures for support are already being implemented at the prefectural and municipal levels.

As a component of its agricultural policy which emphasizes environmental conservation, the Japanese government has established "agricultural and environmental norms" as something that must be observed by farmers covered by policy measures. On top of that, the government is pushing "measures to improve the conservation of farmland, water, and the environment." As part of that effort, the government pays farmers a certain amount in direct assistance under certain conditions that include, as a rule, a minimum 50% reduction from the usual amounts of chemical fertilizers and chemical pesticides used, a consolidated effort within a certain geographical area, and that the assistance recipient be an Eco-farmer as defined under the Sustainable Farming Law. This represents a measure of progress in agricultural policy that considers environmental conservation important.

Chapter 14

From the Production of Conservation Livestock Products to the Hospital Ward

Tomiharu Manda

Eating too much grain kills herbivores

In the days when cattle were not closely watched, there were often accidents in which cattle that had strayed from fenced areas gorged themselves on grain and suddenly died. Dairy farmers therefore all know that you must not give cows too much grain. However, imported grain feed is cheap. If you give grain feed to cattle, milk cows will yield more milk, and fattening cattle will yield marbled meat, increasing one's income. Thanks to the large amount of knowledge gained from research on the first stomach (rumen) of cattle, techniques to feed cattle large quantities of grain have been developed.

From 1984 to 1988, the author worked at a government experiment station in Tokachi, Hokkaido, Japan's foremost dairy farming region. Formerly dairy farmers were encouraged to enlarge their operations, so they increased their herds and endeavored to raise per-head milk yield. But partly because feed provision techniques were underdeveloped many cattle died or became unusable due to diets high in grain feed, and mutual aid association veterinarians were kept very busy treating sick cattle. In light of this experience, veterinarians noticed the importance of maintaining constant rumen fermentation and told farmers about feeding techniques based on the control of normal rumen fermentation. Producers adopted these techniques and, thanks to these efforts, Japan's dairy farming has attained the world's top level for milk yield per head. Regarding beef cattle as well, the domestically produced Japanese black wagyu are heavily grain-fed, and recently a feeding technique has been developed for controlling vitamin A, thereby producing marbled beef that is said to be the most delicious in the world. As this shows, sophisticated production techniques for milk and beef have come about through new knowledge about cattle nutritional science and rumen science. Additionally, progress in livestock improvement technologies based on genetics and breeding theory has realized a quantum leap in the genetic capacity of livestock. Supported by facility automation to save labor and by technological innovation in livestock management using IT, megafarms have appeared throughout Japan, and farms that are run as businesses have eclipsed family farms.

It is well known that the Achilles heel of Japan's livestock industry is the rate of feed self-sufficiency, which is far lower than the rates of other industrialized countries, but whether cattle are given imported grain, roughage, or another kind of feed, rumen science is the basis of livestock nutritional science. Having a good understanding of this principle and applying it in everyday feeding and management makes it possible to raise healthy cattle, and therefore to supply consumers with safe and worry-free products. But if we return to the basics regarding herbivores, there is real meaning in feeding cattle on high-fiber plants instead of grains. From the standpoint of the ruminology of cattle and other ruminants, it is meaningless to raise ruminants as livestock unless this point is exploited. It is now absolutely essential that the Japanese livestock industry raise its rate of feed self-sufficiency. We must "ruminate" on the idea of feeding cattle on high-fiber plants, which only stands to reason.

Without livestock there is no agriculture: herbivores will save the Earth

Increasing the number of cattle and raising their individual capacity inevitably leads to greater demand for grain feed, which has high energy density. For this reason farmers in grain-exporting countries are encouraged to intensify crop farming by the heavy use of chemical fertilizers and pesticides and to convert forest and other land to cropland, which has heavy impacts such as on the environment and biodiversity. Consumers and citizens have few opportunities to perceive this as a problem affecting themselves because agricultural products are traded widely in international markets. Further, the methane and nitrous oxide generated in the digestive tracts of herbivorous livestock and from their waste is an issue as a cause of global warming. But while the methane generated by herbivorous livestock has a deleterious effect on the environment, it is also known that raising such livestock in numbers commensurate with land area has a harmonizing effect on the environment. For example, it is also recognized that grazing on short semi-natural grassland increases biodiversity, and that such grasslands function as an important sink for methane, a global warming gas. Further, it has long been known that the combination of crop and livestock farming is effective in improving the soil's physical and fertility properties through the provision of compost and other organic fertilizer. Thus, the biggest contribution of livestock in this combined system is the improvement of sustained production and land productivity. Although people have tended to overlook this positive assessment of resource use and resource economy by livestock, it is now an important challenge to see how Japanese farmers can build and develop an herbivorous livestock production system, which is effective in conserving the natural environment, around Japan's climate and land features. Especially in Hokkaido, which has the greater portion of Japan's grassland, grazing offers great promise as a means of saving labor and costs. By switching from lot feeding and large herds to intensive grazing techniques, some dairy farmers have managed to raise their income ratios, considerably shrink their huge debts, and save labor. Meanwhile in southwestern Japan, with a climate of high temperatures, high rainfall, and high humidity from the rainy season and into the summer, farmers are working on new farming methods such as a year-round grazing system which takes advantage of flourishing wild grass resources and short grasslands, and systems which combine livestock, rice farming, and forestry. Such initiatives are also important in terms of land conservation, water resource conservation, and conserving the natural resource base including biodiversity. One hope is that in the future even people in small-scale farming and family mixed agriculture will make efforts to further improve the flows of nutrients and energy within their operations and their local areas and achieve sustainable production. Especially in mountainous regions where the topology is complex and populations are aging, building systems that combine livestock, crop farming, and forestry is important not only for conservation of the land and the continuance of communities, but also for supplying the citizens with safe and worry-free food.

Raising the rate of livestock feed self-sufficiency and improving the quality of livestock products

The "Basic Plan for Food, Agriculture and Rural Areas" (below, "Basic Plan") formulated in March 2005 sets a numerical target for raising Japan's overall rate of food self-sufficiency on a calorie basis from 40% in 2003 to 45% in 2015. Raising the rate of livestock feed self-sufficiency is considered crucial to meeting this target, and Japan's livestock industry definitely needs to solve this problem. Among the tasks set forth by the Basic Plan are securing food safety and consumer trust, emphasizing environmental conservation, and facilitating the training of farmers and amassing farmland for their use. Policy measures are being implemented to achieve these goals.

Meanwhile, against the backdrop of this age of gluttony, livestock products are in a transition from emphasis on quantity to emphasis on quality, such as in quality improvement and differentiation. Additionally, the encouragement of localized production and consumption, the rise of the slow food movement, and other such trends seem to feed into the advancing diversification of sales, distribution, and consumption channels for both domestic and imported livestock products.

Recently the public is showing greater interest in the relationship between food and health. In tandem with the advances in food chemistry and analytical chemistry, many new functional components are being discovered in livestock products, for which reason specified health foods and other products are being commercialized in expectation of their beneficial effects. There is no doubt that practical research in this field will realize great advances through closer partnering with the healthcare field.

In June 2005 the Basic Law on Nutritional Education was enacted to promote food education so that people have proper diets. In addition to nutritional education, the Basic Plan includes provisions for encouraging localized production and consumption to link local farmers and consumers. The government also instituted a positive list program relating to standards on the residual pesticides and other substances in foods. In view of these facts, the development of new marketing methods and the like which use labeling to show what functional components were detected in livestock products produced with livestock farming using nature's cycle not only responds to the public's demands regarding food, but is also important to the financial stability of livestock farming operations.

Accommodating the demand for a higher rate of food self-sufficiency

When you say "self-sufficiency," the first thing that pops into the heads of livestock industry people is livestock feed self-sufficiency. But the lack of attention to this matter even among livestock industry people is a sign of our times, in which economic efficiency is regarded as all-important. Generally, "self-sufficiency" reminds older people of the austerities forced upon them during and after World War II. Businessmen on the front lines of international competition and young people who bask in an affluent material culture associate "self-sufficiency" with being anachronistic, closed, feudal, perverse, and other such qualities, so it does not necessarily create a good impression. But recently this term has become a topic of discussion among the public. Lifestyle reassessment is one way this topic has arisen. TV and magazines tell about the agricultural lifestyles in which celebrities and other well-known people move to the country and enjoy nature, and many people now show an interest in homemade and natural food ingredients, and other aspects of agricultural-style self-sufficient living.

Another reason is the emergence of national challenges that are directly connected to the government's policy on food self-sufficiency. An underlying factor is that Japan's self-sufficiency rate is declining year by year, and many people have begun to feel anxious over the heavy dependence of Japan's food supply on other countries. As this shows, although people thought little of self-sufficiency during the years of rapid economic growth, now that many members of the baby boom generation are at retirement age, the importance of self-sufficiency is starting to be understood. Meanwhile, agriculture technology researchers have been handed the very difficult job of raising the self-sufficiency rate. So, to consider what the livestock sector should do to raise the food self-sufficiency rate, let's approach this from the relationship between the food consumption structure and farmland use. In recent years, livestock products have accounted for a larger portion of family food expenditures, but because domestic food production and farmland use are slanted toward rice production, Japan imports a large amount of livestock feed produced on about 4.5 million ha of farmland in other countries. This disharmony between food consumption and farmland use causes problems including the plunder of foreign farmland, domestic livestock environmental problems, excess rice production, and the declining use and dilapidated condition of farmland. In other words, the kinds of food on Japanese dinner tables do not correspond to land use.

To correct this disharmony and form a new cyclical agriculture, it is important that livestock farming be the core of agricultural land − use. Specifically, the challenge is to build a compound agricultural production system that is harmonized through the organic bonding of forested land, cropland (rice paddies), and livestock farming, and whose horizon includes the ocean and rivers.

Marbled beef on festive days

In Japan every year, gala TV programs broadcast on and around New Year's Day feature celebrities smacking their lips over deluxe Japanese beef at well-known hot spring inns and high-class restaurants. Inexpensive imported beef has never appeared on such programs. The fare at low-priced ordinary restaurants uses imported beef, while people eat deluxe Japanese beef on special occasions. This distinction is entrenched in Japanese society. Imported beef is for fast food, which is simple and attractive for its low price, while deluxe Japanese beef is slow food for the rich, and representative of special beef brands, and in particular of localized production and consumption. Since the deregulation of beef imports, domestically produced wagyu has tried to survive by differentiating itself through high-quality beef production, and has managed to produce the finest marbled beef in the world. For the time being, imported beef will not replace marbled beef.

Although the market price of wagyu veal and the price of beef slumped for a while from the influence of BSE, wagyu is now so high-priced that people are calling it a wagyu bubble. Despite the resumption of beef imports, it appears this momentum will continue for a time. However, some people sense danger in the simplification of "marbled beef" (single-item production) everywhere in Japan.

From the environment-friendly production of livestock products to the hospital ward

The public has a very great interest in safe and worry-free food, and the recent orientation toward health is spurring that interest. Owing to the situation described above, the Kitasato University Field Science Center Yakumo Experimental Farm, located in southern Hokkaido, stopped using imported feed in 1994, and is producing beef using feed supplied entirely from the farm (grazing in summer, stored feed in winter). The idea behind this goal is "building a cyclical community that conserves nature and food, and maintains human health." Here we are not bent on pure wagyu breeds. We have created cross-bred cattle that are healthy even in the severe natural conditions of our cold mountainous and hilly region, and endeavoring to produce safe, worry-free beef. Of course the beef is not marbled, but is red meat with little marbling that can be purchased at low prices. This would certainly not be economically sustainable through the usual distribution channels. Fortunately the Greater Tokyo Area Consumer Cooperative took the leap and decided to purchase our beef. Understanding of our efforts has gradually increased, and now our beef is being used in many places including school lunches, the meals at local well-known hot spring inns, and restaurants.

The director of the Nutrition Department at the Kitasato University Hospital visited the farm himself. Healthy grass grows from pasture soil on which we apply organic fertilizer but no pesticides. When the cattle eat this grass, they change it into safe, delicious red meat through the action of microorganisms in the first stomach, an important digestive organ in herbivores. The hospital saw our unforced production system, which uses local resources such as water, soil, grass, and air, and noted the safety and soundness of our beef, and started using it in patient meals last autumn. The patients say it is delicious.

In other words, here we can see the partnering between "agriculture and the environment and medicine." Additionally, we host the students of Kitasato University extension lectures and give an understanding of our system to citizens as university customers. Although there are still few instances of practice like this, we believe that our efforts to fully manifest the physiological characteristics of herbivorous livestock and the ability to sustainably use natural resources to produce livestock products based on the original thinking behind livestock farming reflects consumers' expectations. Our production system at Yakumo Experimental Farm has been chosen as a topic of research commissioned by the government starting in FY2006, which means that testing will be conducted throughout our local area, Yakumo Town. Our production-to-dinner table endeavor for Kitasato Yakumo cattle developed in this northern clime is about to expand from the university farm to the region as a practical version of "agriculture, environment, and medicine."

Part III
A Look at Avian Influenza from the Perspective of Agricultural Environment and Medicine

15 A Message from the Symposium Organizer ································ 71
    Tadayoshi Shiba

**16 Looking at Bird Flu from the Perspective of Agriculture,
     Environment and Medicine** ·· 73
    Katsu Minami and Shinji Takai

17 Current Status and Issues of Zoonotic Viral Diseases ·················· 79
    Yasuhiro Yoshikawa

18 Highly Pathogenic AIV Infection and Countermeasures ················ 85
    Shigeo Yamaguchi

19 Wild Bird Migration and Behavior in Relation to AIV Infection ········· 89
    Yutaka Kanai

20 Infection of Wild Birds and Current Status ································ 95
    Kumiko Yoneda

**21 The Threat Posed by New Types of Influenza:
     AIV and its Impact on Humans** ··· 98
    Nobuhiko Okabe

22 Highly Pathogenic AIV and Vaccination Measures ······················ 101
    Tetsuo Nakayama

Chapter 15

A Message from the Symposium Organizer

Tadayoshi Shiba

Allow me to make a few remarks as representative of the organizer in relation to Kitasato University's 3rd Agromedicine Symposium.

In 2006, the Science Council of Japan was restructured through consolidation of its 7 earlier divisions — Literature, Law, Economics, Physical Sciences, Engineering, Agriculture, and Medicine — into the 3 divisions of Humanities, Life Sciences, and Physical Sciences and Engineering. In its previous form, the Science Council was supported by the research liaison committees and academic conference activities of the many academic societies that comprised its backbone, but since its reorganization, such society-based activities have declined, giving way to issues-based activities that span a range of fields and bring together experts in specialized areas of different disciplines.

This kind of change has had a significant impact on academic society activities, allocation of research funds, and research trends in each field. For example, the Science Council has from its 20th term organized its committees according to issues and is putting greater emphasis on the contribution that academia can make to society.

The Science Council of Japan is the cornerstone of academia in Japan, and I think that all universities and research institutes need to pay attention to the direction taken by the Council with respect to education and research in Japan. The changing trend of academic conferences is no doubt being driven by a reconsideration of the path taken by science here and a desire to explore new fields, and I think that the initiatives taken in the Council's 20th term give much reason for hope for the future.

To move on to another subject, human activity has had the effect of disturbing the natural order of things on Earth. As a result, ecosystems are beginning to be undermined by a myriad of little invaders, from viruses to weeds.

In September 2001, Japan suffered its first outbreak of BSE (bovine spongiform encephalopathy), a disease that opens holes in the brain tissues of cows and damages the central nervous system. The consumption of parts of cows infected with BSE, particularly the most risky parts such as brain and other nervous tissues and distal ileum, can, albeit very rarely, lead to human infection, causing senility and death. This is just one more little invader.

And now there is a new source of worry. The avian influenza that has broken out here and there in Asia over the past few years has spread to Europe too. The pathogen, a virus, tends to mutate with successive outbreaks, and there is a growing chance of a new influenza virus appearing that will spread like wildfire through human society. This is another case of bioinvasion.

The WHO and many countries have started preparations to guard against such an eventuality, and Japan is no exception. The Ministry of Health, Labour and Welfare on November 14, 2005, announced an action plan that includes its declaration of a state of emergency in the event of an outbreak of a new influenza virus in Japan.

Agriculture is human activity applied to the environment. There is no human activity that does not affect the ecosystem, which could be seen as a huge symphony of life forms. The order of the ecosystem created by the countless organisms within the environment could also be seen as a network of innumerable — and largely invisible — interactions between life forms and resources within the environment. This network and the organisms that comprise it automatically restore equilibrium when that equilibrium is disturbed, and as a result, the natural world can maintain this equilibrium forever.

Unless we seriously tackle such issues as BSE, the current bird flu problem, and other problems caused by these little invaders that are bound to crop up in the future, the human race faces a very dark future. These little invader issues are intimately related to agriculture, environment, and medicine, and cannot be resolved unless we consider agriculture, environment, and medicine as one.

The Science Council of Japan's establishment of a new Life Sciences division is in my mind a vital approach to resolving these little invader issues. Our own new educational and research goal at Kitasato University — the integration of agriculture, environment, and medicine — is very much the same kind of approach. This is an area in which the emphasis is on surmounting disjunction between different areas of knowledge.

It is my sincere wish that holding this Kitasato University 3rd Agromedicine Symposium "A Look at Avian Influenza from the Perspective of Agriculture, Environment, and Medicine" with participation from experts in all three fields will help contribute to the resolution of the issue of bird flu. I hope that the event will be a forum for meaningful and pragmatic discussion that gives rise to new ideas for integrating agriculture, environment, and medicine. I would like to end by expressing my heartfelt thanks to the professors who so readily agreed to speak at this symposium.

Chapter 16

Looking at Bird Flu from the Perspective of Agriculture, Environment and Medicine

Katsu Minami
Shinji Takai

Little invaders

There is a book titled *Life Out of Bounds* (1998) written by Chris Bright, an environmental journalist and magazine editor, while serving as the chief editor of World Watch, a magazine issued by the Worldwatch Institute. The Japanese language edition, titled *Little Invaders that Destroy the Ecosystem*, was published in 1999 by Ie‐no‐Hikari Association.

In his acknowledgments at the front of the book, the author writes that he co‐hosted a Worldwatch press conference on bioinvasion — "how human activity is 'stirring up' the Earth's organisms — whether viruses, weeds or whatever?and about why the consequent levels of biotic mixing tend to injure our societies and the natural world. The press conference was our way of inviting the public to explore this terrain with us. *Life Out of Bounds* is a continuation of that process."

Worldwatch founder Lester Brown too writes as follows in a foreword that he contributed to the Japanese edition:

"This book that you are about to read was written with the aim of providing you with a clear picture of how these life forms travel, and the ecological damage they cause. This damage is brought about through the invasion of introduced species that spread at a prolific rate. Introduced species are life forms that have been introduced to ecosystems other than the ecosystem in which they themselves evolved. Once an introduced species gains a foothold in a new location, there is a chance that it will proliferate rapidly and, in the competition for the resources required for survival, overwhelm native species and impede their propagation. If the introduced species is a microorganism, it may trigger an epidemic, and if it is a predator, it may predate on native species to an extent that it eradicates them.

"How can we maintain a balance between trade and the need to protect the world's ecosystems? This is the core issue tackled by this book. The health of our economies clearly depends on a high level of international trade, but it is equally clear that the health of our ecosystems depends on how well we can keep most of this planet's life forms within their native habitats. I think that ensuring a balance between these two needs will become one of the key environmental issues of the new century."

In September 2001, Japan suffered its first outbreak of BSE, a disease that opens holes in the brain tissues of cows and damages the central nervous system. The consumption of parts of cows infected with BSE, particularly the most risky parts such as brain and other nervous tissues and distal ileum, can, albeit very rarely, lead to human infection, causing senility and death. This is just one more little invader.

And now there is a new source of worry, though one of a kind that the world has experienced several times in the past. The avian influenza virus (AIV) that has broken out here and there in Asia over the past few years has emerged in Europe too. This year too, the highly pathogenic H5N1 strain of AIV has emerged in the province of South Chungcheong in Korea, and in the prefectures of Miyazaki and Okayama in Japan. In Korea, 19 H5N1 outbreaks were recorded between December 2003 and March 2004 in poultry farms and other locations, resulting in the death or extermination of 5.3 million chickens and ducks. Four workers involved in the extermination work were also infected by the virus.

Viruses like this tend to mutate with successive outbreaks, and there is a growing chance of a new influenza virus appearing that will spread like wildfire through human society. This is another case of bioinvasion.

The WHO and many countries have launched preparations to guard against such an eventuality, and Japan is no exception. The Ministry of Health, Labour and Welfare in November 2005, announced an action plan that includes its declaration of a state of emergency in the event of an outbreak of a new influenza virus in Japan. This action plan is based on guidelines provided by the WHO, and delineates measures to be taken in the event of both overseas (A) and domestic (B) outbreaks according to 6 phases from interpandemic (normal) to pandemic phases.

Agriculture is human activity applied to the environment. There is no human activity that does not affect the environment or ecosystem, which could be seen as a huge symphony of life forms. AIV or BSE could be seen as evidence that human activity is undermining the natural order and network of interaction created by countless life forms in a diversity of environments.

We need to seriously tackle such issues as BSE, the current bird flu problem, and other new and reemerging infectious diseases that are bound to crop up in the future, considering them in relation to the environment and ecosystem. These issues are intimately related to agriculture, environment, and medicine, and cannot be resolved unless we consider agriculture, environment and medicine as one.

Recent spread of AIV

The risk of highly pathogenic (HP) AIV H5N1 spreading throughout the world has grown since 2004. Up to then, infection had been limited to birds in Asia alone, but from 2004, it began to spread to Europe and Africa. While everyone now recognizes the importance of international pandemic countermeasures, such countermeasures are still imperfect in Asia, and outbreaks cropped up in both Korea and Japan at the end of last year and the beginning of this year.

Outside of East Asia, HPAIV H5N1 has also been identified in Azerbaijan, Djibouti, Egypt, Iraq, Iran, Greece, Turkey, Romania, Kuwait, Russia, Mongolia, Kazakhstan, Tibet, Croatia, Italy, Macedonia, Canada, Sweden, the UK, and the USA, and has been found to infect a wide range of birds, including domestic and wild ducks, turkeys, swans, parrots, chickens, flamingos, and falcons.

Cases of AIV H5N1 infection of humans have occurred in the following 10 countries: Azerbaijan, Cambodia, China, Djibouti, Egypt, Indonesia, Iraq, Thailand, Turkey, Nigeria and Vietnam.

Numbers of people infected (deaths in brackets) between December 2003 and February 6, 2007 are as follows: Vietnam: 93 (42), Indonesia: 81 (63), Thailand: 25 (17), China: 22 (14), Egypt: 19 (12), Turkey: 12 (4), Azerbaijan: 8 (5), Cambodia: 6 (6), Iraq: 3 (2), Nigeria: 1 (1), Djibouti: 1 (0). This makes for a total of 271 persons infected and 166 deaths. The above figures are for cases reported and confirmed, and actual infections and deaths are likely to be higher. (source: Infectious Disease Surveillance Center, National Institute of Infectious Diseases)

Avian influenza epidemics

Pandemics of Spanish flu and other strains of influenza have swept the world several times in the past, causing large numbers of deaths. The following is a brief chronology of such epidemics.

- 1889–90: Isolation of "influenza germ" from an epidemic, but it was not the real pathogen, which was an H2N2 strain.
- 1918–19: Spanish flu infects 500 million people and kills 1% of the world's population. H1N1 subtype. In Japan, 23 million infected, 390,000 deaths, though higher estimates also exist.
- 1933: Type A influenza isolated. Type B isolated in 1940, and type C in 1947
- 1943: Vaccines for types A and B influenza developed in the USA using embryonated eggs
- 1946: Italian flu, an H1N1 subtype that went pandemic in Europe
- 1948: WHO World Influenza Centre established in London
- 1951: Prototype vaccine developed in Japan. Mass production from 1962
- 1957–58: Asian flu pandemic. H2N2 subtype. In Japan, 980,000 infected, 7,000 deaths
- 1968: Hong Kong flu pandemic. H3N2 subtype. In Japan, 140,000 infected, 2,000 deaths
- 1977: Russian flu. H1N1 subtype
- 1997: Hong Kong H5N1 outbreak in domestic animals. High fatalities in infected animals
- 2003–04: H5N1 pandemics among birds in various Asian countries. Human deaths in Thailand and Vietnam

2005 - :   H5N1 pandemic among wildfowl inhabiting Qinghai Lake, China. Later spreads to Europe and Africa.

AIV human infection numbers

An incident occurred in 1997 that forewarned of a global pandemic. At the time, influenza was spreading among chickens and other domestic fowl in Hong Kong, and the causative AIV infected 18 humans, resulting in 6 deaths. This incident shocked the world for the fact that AIV, up to then thought incapable of infecting humans, made a direct jump from domestic fowl to humans, and that it was moreover a highly pathogenic (HP) strain (type A influenza H5N1 subtype).

AIV infection of humans has been reported every year since then, with deaths increasing exponentially every year, as follows:

1997 :   H5N1 China (Hong Kong): 18 infected (6 deaths)
1998 :   H9N2 China: 9
1999 :   H9N2 China (Hong Kong): 2
2003 :   H7N7 Holland: 89 (1)
2003 :   H9N2 China: 1
2004 :   H7N3 Canada: 2
2004 :   H10N7 Egypt: 2
2003 - 07 : H5N1

As of February 6, 2007, there have been 271 humans infected (166 deaths) in 10 countries. Deaths per country are provided above.

Waterfowl are the origin of AIV

AIV has been isolated from over 90 wild bird species so far, a figure that suggests that almost all avians are susceptible to AIV infection.

In the 1970s, AIV was found to be carried by a high percentage of waterfowl around the world, particularly in healthy migrants such as wild ducks, wild geese, swans, curlews, plovers, seagulls, and terns. Wild ducks in particular show a very high percentage of AIV, and wild waterfowl are accordingly now regarded as the natural hosts of the virus in the wild.

Waterfowl carrying AIV at a particularly high level include young mallards just prior to migration in late summer, and other duck species inhabiting ponds and rivers in early winter. A great many waterfowl migrate annually to Japan from Siberia and China, and one study found that 3% of those migrants carry AIV. Curlews and plovers have also been found to carry AIV.

It was shown in the 1970s that a hybrid AIV could be generated by infecting pigs simultaneously with AIV and human influenza. After the discovery that AIV originates in waterfowl such as ducks and swans, the sequencing in the 1980s of the genes of various types of influenza virus proved that wild waterfowl are the natural hosts of AIV and that the original strain spread to various other animals as it mutated.

Impacts of human activity

Human activity has brought about many major changes in the natural world. Such impacts are permissible when kept within limits that enable the ecosystem to recover fairly easily, but we appear to have overstepped those limits.

Up to now, AIV had remained isolated in nature within waterfowl, curlews, and plovers, but owing to international trade, new forms of culture, industrialization of poultry farming and so forth, the virus's ecosystem, distribution, range of hosts, and virulence has changed dramatically. Causative factors include the international trade in pet birds, domestic waterfowl farming, free-range poultry farms, sale of live birds, ornamental and fighting cock trade, and increasing size of poultry farms.

Ducks and geese, which tend to be raised in the open, show a higher prevalence of AIV than chickens do. Because the ponds required for raising waterfowl are normally located outdoors, they are prone to contamination by migrating waterfowl and other wild birds flying overhead or landing on them.

It was previously thought that AIV would crop up only very rarely in the kind of intensive poultry farming practiced in developed countries, but such is no longer the case. Once the virus invades such an intensive environment, it can spread rapidly to affiliated farms, resulting in much more extensive damage. This issue is now deservedly attracting serious consideration.

Countering new strains of influenza

What can we do to prevent the appearance of a new and highly pathogenic H5N1 influenza virus? And what kind of countermeasures do we need to take should such a virus appear? What about choice of vaccine strains, or the production of drugs that fight the influenza virus?

Now that we are already at Phase 3 in the WHO's six public health risk phases and are considering preparedness for Phase 4, what kind of measures should we be thinking about from the perspective of agriculture, environment, and medicine? It is my aim to make this a meaningful symposium that integrates agriculture, environment, and medicine through having experts from various specialized fields provide their views on this subject.

There are of course many related issues, such as national crisis management above and beyond medicine, the issue of balancing crisis management with individual freedom, the role of the media, risk assessment and risk communication issues, household preparedness, and so forth, and I hope that this symposium will also provide opportunities for considering such issues.

Closing comments

It is my fervent hope that a new influenza virus does not emerge, but to such an end I think we need to once more go over the words of Francis Bacon or Rachel Carson and reconsider our relationship with technology and nature.

The ecosystem could be seen as a huge symphony of life forms. The order (or maybe "harmony" would be more appropriate) of the ecosystem created by countless organisms within the environment could also be seen as a network of innumerable — and largely invisible — interactions between life forms and resources within the environment. This network and the organisms that comprise it are designed to automatically restore equilibrium when that equilibrium is disturbed, and as a result, the natural world can maintain this equilibrium forever. However, this "forever" may well now be finite.

Insofar as each and every organism seeks to preserve its line as it evolves, the fight for survival knows no end. The AIV issue is an extension of this struggle, and as such, we need to apply whatever human wisdom we have to prevent and to prepare for outbreaks. Tackling this issue as soon as possible is one way of ensuring a brighter future for humanity.

Chapter 17

Current Status and Issues of Zoonotic Viral Diseases

Yasuhiro Yoshikawa

Opening comments

Human beings are heterotrophic organisms that depend on animals and plants as sources of nourishment. Most of our needs for protein and fat are now met by consumption of the milk, meat, internal organs and other parts of domestic animals. We have had a long relationship with domestic animals, some of which were already living among us when we started farming the land 10,000 years ago. A look at that history shows that almost all current infectious diseases suffered by humans have animal origins. In other words, diseases such as smallpox, measles, and influenza that were once thought to be unique to humans, all pathogens originate in other animals or share common ancestors with viruses infecting other animals. There are also many infectious diseases even today that can be passed between people and domestic animals. We humans do not inhabit a special world separate from that of other animals.

From animals to humans

Zoonotic diseases are diseases caused by a pathogen that infects both animals and humans (but natural hosts infected by the pathogen often do not suffer any adverse effects). They consist mostly of diseases passed on to humans from animals, and diseases originally passed on to animals from humans and then back to humans from the infected animals (so-called recurrent infections, e.g. dysentery, tuberculosis, viral hepatitis, and other diseases found in monkeys).

Zoonotic diseases include such well-known examples from ancient times as plague, which is transferred from wild rodents (rats, etc.) to humans through fleas (and is by no means a disease of the past, still being prevalent in the continents of Africa, Asia, and America), and rabies, which is passed on to humans from infected dogs, bats, and other animals. There are of course many other parasitical, rickettsial and chlamydial, bacterial, and viral diseases affecting humans. In 1959, a WHO and FAO joint expert committee listed over 150 such diseases, and now there are thought to be 500-700 noteworthy diseases. Infectious diseases that have sent shockwaves throughout the world in recent times include diseases of wild animal origin such as Ebola hemorrhagic fever (HF), Nipah virus infection, SARS, and West Nile fever; diseases of domestic animal origin such as O-157, BSE, and HPAIV; and diseases of arthropod origin such as dengue fever, dengue hemorrhagic fever, and malaria. About two-thirds of all viral diseases to have emerged in the latter half of the 20th century are zoonotic. Infectious diseases of domestic animal origin such as salmonella, hepatitis E, O-157, and BSE warrant serious consideration also from the food safety perspective since they invariably spread through foodstuffs.

Retrospectively, it was in 1980 that the WHO declared that smallpox had been eradicated. Though it is only one pathogen, this was the first time in history that mankind had defeated a virus (though recently people have voiced concern that it has not been completely eradicated ironically insofar as it continues to exist in the form of samples that might some day be used as pathogens in acts of bioterrorism). With the development of antibiotics, we also became able to suppress bacterial infections, giving rise to optimism about our ability to protect ourselves from infectious diseases. In Japan too, the infectious diseases that were long the top causes of death declined rapidly after the 2nd World war, making way for cancer to become the No.1 cause of death by 1950. As circulatory disorders became the 2nd most prominent cause of death, Japan's healthcare authorities began to focus more on welfare and countering cancer and lifestyle diseases rather than infectious diseases.

However, new infectious diseases such as AIDS and various viral hemorrhagic fevers have emerged worldwide, and diseases such as dengue fever and tuberculosis have reemerged to become serious threats to human health once again. Excessive use of antibiotics has given rise to the spread within hospitals of antibiotic-resistant bacteria such as MRSA, VRE, and VRSA. Given such developments, the WHO has revised its optimistic forecasts regarding the fight against infectious diseases, and countries throughout the world have declared states of crisis with regard to infectious diseases.

Factors behind the occurrence and spread of zoonotic diseases

Most zoonotic diseases can be traced to developing countries. The reasons for this include increased contact with pathogens carried by wild animals in tropical rainforest and other natural habitats during development of human production activities (Ebola HF, Marburg disease, monkeypox), disturbance of ecosystems by rodents and other animals whose numbers have been elevated by increased human productivity (Bolivian HF, Lassa fever, Argentine HF, etc.), establishment of infectious disease in cities of developing countries, which is normally circulated between monkeys and mosquitoes in forests owing to rapid urbanization and population concentration combined with poor urban infrastructure (yellow fever, dengue fever, dengue HF, etc.), and rapid spread of infection from developing to developed countries as a result of the rapid air transport of both people and animals (Lassa fever, Marburg disease, SARS).

There are also contributing factors in developed countries, such as the keeping of wild animals as so-called exotic pets (tularemia, plague, monkeypox, etc. transmitted by pet prairie dogs), and contact with wild animals during outdoor recreation such as camping or forest walking (Japanese spotted fever, scrub typhus, Hantavirus pulmonary syndrome and Lyme disease transmitted by such animals as wild rodents and ticks, echinococcosis transmitted by foxes, etc.). New infectious diseases have also emerged in developed countries as a result of the pursuit of economic efficiencies in the form of intensive factory farming and rendering of animal parts as sources of protein (salmonella, BSE, O-157, etc.). In recent years, moreover, we are seeing transmission patterns of a more complicated kind, such as the Hendra and Nipah viruses transmitted from tropical fruit bats — up to now not known to be carriers of pathogens — to humans through domestic animals.

The chances of coming into contact with infectious diseases in humans transmitted by domestic animals such as pigs (Nipah virus), horses (Hendra virus), cattle (BSE), or chickens (HPAIV) are much higher than for those of wild animal origin. Domestic animals are increasingly raised for human consumption in large-scale factory farms, and once a pathogen invades such an intensive rearing environment, it can spread like wildfire, with the likelihood that its frequent transmission among hosts in such an environment will also facilitate genetic mutation, making for a much more dangerous situation than in the past.

Even among wild animals, we might be facing new risks. For example, increasing environmental pollution might reduce host immune functions, as a result of which a virus that has up to now coexisted with a host suddenly begins to spread explosively (North Sea seal virus, etc.), or environmental pollutants might elevate the frequency of virus mutation, because they were frequently mutagenic chemical substances. This kind of possibility suggests a need for conception change and actions different from earlier measures for suppressing zoonoses and avoiding risks. Conservation medicine (http://en............) is a new approach to the control of zoonoses that incorporates the concept of environmental conservation in the consideration human and animal health.

Warning to humanity

The way in which zoonoses emerge and spread is changing in connection with the expansion of human production activities, pursuit of economic efficiency, changing lifestyles, and so forth. In this respect, zoonoses have much in common with environmental pollutants such as PCB, DDT, and dioxins. There is nothing evil about pursuing comfort and convenience, but if in our anthropocentric pursuit of ever more advanced technology we continue to ignore the need for balance and continue to destroy the environment and ecosystem, we are doomed to suffer the consequences. Attempts to resolve issues by pushing the contradictions of developed countries onto developing countries or by a country just looking out for itself are already proving to be bankrupt. What is needed is global cooperation between governments on countermeasures to zoonoses led by the WHO and OIE. National governments should also be remind that to avoid covering up or failing to report outbreaks, or all clear declarations under issuing premature. Other acts aimed simply at protecting one's own country's economy or calming the populace will in the end only raise the risks of a global infection (SARS in China, HPAIV in Southeast Asia, BSE in the UK, etc.).

Even the USA, which has the most advanced infectious disease defense system in the world and is home to the Centers for Disease Control and Prevention (CDC) that plays a leading role in controlling infectious diseases worldwide, has not had an easy time controlling zoonoses like West Nile fever that are transmitted through wild animals (birds and mosquitoes). West Nile fever first appeared in eastern New York in 1999, infecting 7 people, but by 2003, it had spread throughout the country and still shows no signs of abating, with infections now standing at over 8,000 and deaths at over 200. The USA is also finding it extremely difficult to suppress plague endemic to arid Midwest regions (being transmitted between prairie dogs and fleas) and rabies transmitted by bats.

Meanwhile, the fact that SARS, which is thought to be of wild animal origin, spread throughout the world in a matter of months demonstrates that national borders and other artificial barriers are no obstacle to modern infectious diseases. HPAIV H5N1, the subject of this symposium, has also spread from Asia to the Middle East, Europe, and Africa. The number of countries affected, the scale of infection, and virulence that has enabled it to directly infect not only pigs but also humans, has prompted the WHO to issue dire warnings about the dangers it poses. In addition to conventional downstream, end-result-oriented infection countermeasures targeting people and animals (Ministry of Agriculture, Forestry and Fisheries [MAFF], Ministry of Health, Labour and Welfare [MHLW]), in the 21st century, zoonoses originating in wild animals need to be investigated from a more upstream perspective that also considers the environment and the ecology of pathogens parasitizing wild animals and natural hosts in order to develop more global countermeasures.

The path to controlling zoonoses

Including pathogenic microorganisms, there are currently about 1.4 million known species on Earth (approximately 750,000 insects, 280,000 other animals, 250,000 higher plants, 70,000 fungi, 30,000 protozoans, 5,000 bacteria, and 1,000 viruses). When one considers the complexity of the ecosystem that these organisms have built up as the present-day descendants of 3.7 billion years of life on Earth, it is impossible for we humans to completely control zoonoses for the sake of our own convenience. Basically we need to recognize the importance of biodiversity and seek to achieve a balanced coexistence with other life forms.

Even so, we need to do what we can to control infectious diseases that endanger humanity. The organizations charged with the responsibility of controlling infectious diseases on an international level are the Geneva-based WHO for human infectious diseases, and the OIE, headquartered in France, for animal infectious diseases and infectious diseases whose origins can be traced to foodstuffs. Because OIE decisions frequently directly affect domestic animals in various countries and trade in foodstuffs of domestic animal origin, the OIE also serves as an affiliate of the WTO.

The expert committees of these international organizations frequently use risk analysis as an analytical method. This methodology was originally used to decide international safety criteria with respect to humans for drugs, food additives, and so forth, but has come to be used also in the control of food poisoning and infection by microorganisms. Risk analysis is a field that merges natural science with social science, and is made up of three key aspects —

risk assessment, risk management, and risk communication. Based on a scientific, quantitative risk assessment, the parties concerned (risk managers) consider cost-effectiveness and draft a realistic plan that they explain to others in easily understandable terms, and attempt to establish a more efficient defense system. In Japan after the BSE panic, the Food Safety Commission was established within the Cabinet Office as a risk assessment organ independent from risk management organs. International organizations are already bringing together infectious disease experts and government officials from different countries or regions in field-specific forums to consider measures for the sustained control of infectious diseases.

However, the control of such diseases is basically a political and economic issue. As long as poverty, famine, and war continue, there is little hope for improving public hygiene globally. The path to controlling infectious diseases is one of international cooperation in the building of standards and systems for global defense against such diseases that also respect diversity in the form of national and regional differences in culture, national character, and everyday life and customs.

Japan's new zoonosis countermeasures

After the postwar period of rapid economic growth, dramatic changes in the social system and values fueled the trend towards nuclear families and declining birthrate, and pets as companion animals came to serve as substitutes for people. Then during the economic bubble of the 1980s, in place of the traditional species of pet animals, the import and keeping of exotic animals became very popular. Japan's birthrate declined and population aged at a pace that was exceptional even among the developed countries, and Japan also stood out from the rest in the quantity of its wild animal imports. These changes in society and diversification in lifestyles prompted increasing concern over the possibility that novel zoonoses would emerge, and so when the Infectious Diseases Control Law was enacted (effective from 1999), in addition to diseases transmitted between people, zoonoses too were considered for the first time, and with an expansion of the Rabies Prevention Law, cats, skunks, raccoons, and foxes in addition to dogs became subject to legal quarantine, as did monkeys. However, other infectious diseases and animal species were not subject to regulation, and so when the Infectious Diseases Control Law came up for revision 5 years later, stronger measures were considered.

For this revision, data on infectious diseases, the realities of imported animals, and disease risk assessment was obtained and analyzed. An MHLW zoonotic disease study team carried out a first-ever zoonosis risk analysis. As a result, a total import ban was imposed on all Chiroptera (bats) and rodents of the Mastomys genus (the natural hosts of Lassa fever) from November 2003, and requirements such as import notification, health certificates, and tethering according to risk level were applied to all other animals apart from prairie dogs and civet cats whose import was already prohibited, and monkeys and carnivores already subject to legal quarantine. In other words, unlike previous revisions which tended to simply increase animal quarantine, the new revision applied import bans to certain species, tethering orders, stronger measures against introduced animals and indigenous wildlife (migratory birds, crows, etc.) including surveillance systems, investigation of animals in the event of outbreak of a zoonosis, and stronger measures to combat zoonoses. Particularly the animal import notification system and requirements for health certificates and furnishing of proof of non-infection with certain pathogens effectively put a stop to the import of wild animals that had gone unchecked up to then, and this has proved to be an effective alternative to quarantine as a means of avoiding risks.

With respect to wild and domestic animals within Japan, everyday surveillance is vital, which means that it is also vital to establish an organization for diagnosing infections in animals. With regard to high-risk infectious diseases, there is a need to identify high-risk localities, localities in which animal intrusion is likely, and habitats of wild animals carrying the infectious diseases concerned, and take comprehensive measures to combat the spread of the disease, curb the number and habitats of natural hosts and animal vectors, exterminate intruders, and so forth. This is a field that calls for cooperation between central and local government, between MAFF and the MHLW, and between doctors and veterinarians.

Chapter 18

Highly Pathogenic AIV Infection and Countermeasures

Shigeo Yamaguchi

Opening comments

Avian influenza (AI) is a disease caused by a type A influenza virus infection. It is one of the most serious infectious diseases of the livestock industry because of certain strains of this virus result in the death of the majority of birds in chicken or turkey poultry farms. AIV was previously thought not to infect and cause illness among humans, but in 1997 in Hong Kong, the H5N1 subtype of highly pathogenic avian influenza (HPAI) virus jump to humans, and killing 6 out of 18 infection. HPAI has been recognized as a important zoonosis ever since this shocking event.

In this symposium, I want to explain the importance of HPAI and the control measures from the agricultural perspective, and also want to reffer to countermeasures to new pandemic of human influenza.

Emergence of HPAI

AIV is categolized into two types, low pathogenic and high pathogenic, by the pathogenicity to chickens.

In nature, low pathogenic AIV is found at high levels mainly in waterfowl, and almost all AIV found in the wild birds is of the low pathogenic type. However, as they jump from waterfawl to poultry, a low pathogenic AI viruses mutate into high pathogenic strains that cause 100% fatalities among chickens.

Wild ducks (Anseriformes) harbor low pathogenic AIV at a high level, excreting the virus in feces as they migrate. This virus has an affinity for ducks and does not readily infect other species (orders) such as chickens, turkeys, and other poultry (order Galliformes). However, it can infect such species at a very low rate, and even more rarely establishes itself as a low pathogenic AIV that adopt to poultry and changed to show low infectivity to ducks.

If low pathogenic AIV that has established itself in chickens infects other chickens, certain strains can mutate into high pathogenic AIV capable of causing 100% fatalities in infected chickens. For some reason, up to now low pathogenic AIV that has mutated into high pathogenic AIV has all been of either H5 or H7 subtypes, suggesting that these subtypes are particularly adept at mutating into high pathogenic AIV in chickens.

Because AIV shows the above character, MAFF has defined "highly pathogenic AIV" (HPAIV) for not only high pathogenic AIV infection but also all H5 or H7 subtype infection of even low pathogenic AIV and establishes a movement control zone around affected farms.

HPAIV outbreaks in poultry

1) Outbreaks of low pathogenic types

As stated above, MAFF has classified even low pathogenic AIVs of the H5 and H7 subtypes as HPAIV, labeling them "low pathogenic type" to distinguish them from high pathogenic AIVs.

In May 2005, a type A influenza virus of H5N2 subtype was isolated from specimens collected to investigate a decline in laying at an egg laying farm in Ibaraki Prefecture, Japan. In an experiment to test the pathogenicity of the isolated virus on chickens through intravenous inoculation it was found to be a low pathogenic virus that produced no symptoms in the tested chickens.

Diagnosing the situation as an outbreak of low pathogenic HPAIV, Ibaraki Prefecture ordered the extermination of the chickens on the affected farm, established a 5-km-radius movement control zone around the farm, and inspected all farms within that radius for the presence of AI. This survey turned up a succession of chicken flocks showing no symptoms but possessing antibodies or showing up positive in virus isolation tests. In the end, 40 farms in Ibaraki Prefecture and 1 in Saitama Prefecture proved to be positive, and 5.78 million birds were destroyed before the control order was finally lifted in April 2006.

The isolated virus was a low pathogenic type that had adapted to chickens. It was shown to be highly infective in chickens, but failed in experiments to infect ducks.

2) Outbreaks of high pathogenic types

Between the latter half of 2003 and the end of 2004, outbreaks of HPAI by H5N1 subtype were reported in 10 Asian countries. These outbreaks reportedly resulted in the death or extermination of over 100 million fowl. In Japan, outbreaks occurred on 4 farms in Yamaguchi, Oita, and Kyoto Prefectures, leading to the death or extermination of about 300,000 fowl before all control orders were finally lifted in April 2004.

The isolated virus was a high pathogenic type that caused the death of all 8 test birds within 1 day of intravenous inoculation and within 3 days of nasal inoculation. Gene sequencing showed that all strains isolated in Japan were closely related with each other and with Korean strains, but differed in lineage from those of Thailand, Vietnam, Indonesia, and other Southeast Asian strains, showing that the outbreaks in Japan and Korea were caused by closely related viruses.

Three years later, in January 2007, Japan suffered another outbreak of HPAI by the H5N1 subtype, occurring up to now on 3 farms in Miyazaki Prefecture and 1 in Okayama Prefecture.

The isolated virus is a high pathogenic type that caused the death of all test birds within 1 day of intravenous inoculation. Gene sequencing showed that the strains isolated on each farm were all closely related and belonged to a lineage isolated in 2005 from geese on

Qinghai Lake in northern China. Reported to have caused infection and human deaths over an increasing number of regions throughout the world in recent years, this AI virus lineage has been isolated from waterfowl, domestic fowl, and humans who died from infection in 2005?2006 in Mongolia, Russia, the Middle East, Europe, Africa, and Korea.

In Japan, only one-off outbreaks have occurred, with no cases so far of infection spreading beyond the farm concerned in each outbreak. This minimum damage is considered at the present time to be the result of early discovery and response.

Measures for controlling HPAI

1) Disease control guidelines

HPAI control in Japan is carried out according to MAFF's Guidelines for Controlling Specific Livestock Infectious Disease Related to HPAI. The basic aims of these guidelines are to eradicate AI through destruction of infected fowl in accordance with the control policies of affected countries worldwide and to implement measures for preventing the endemic condition. The guidelines accordingly call for the building of stronger surveillance systems for early discovery and crisis management systems for preventing the spread of the disease.

In the event that domestic fowl showing signs of AI infection are discovered on a farm, the discoverer is required by law to notify the local Livestock Hygiene Service Center. On receiving such a notification, the responsible Center staff members immediately carry out an on-site inspection and launch tests to determine the nature of the disease. On receiving the results of virus isolation and other tests carried out by the Livestock Hygiene Service Center concerned, the National Institute of Animal Health (NIAH) conducts tests to determine subtype and pathogenicity. If, as a result of the above tests, the isolated virus is found to be a high pathogenic strain or of H5 or H7 subtypes, MAFF and prefectural authorities announce the outbreak of HPAI, set up a task force, and launch control measures.

Key control measures include the killing and disposal of fowl at the affected farm, disinfection of the farm, isolation of fowl on epidemiologically related farms, and establishment of a zone in principle covering the area within a 10 km radius of the affected farm, within which the movement of domestic fowl and related items is restricted. As the control measures are implemented and the cleanliness of facilities within the movement control zone is confirmed, the control zone area is gradually narrowed down and the control order eventually lifted.

2) Vaccination-based disease control

An AI vaccine for poultry has been developed and used in several countries. Because low pathogenic H5 and H7 subtypes may mutate into high pathogenic strains with repeated infection of fowl, the low pathogenic live vaccine is not recommended, and vaccines approved up to now around the world have been either inactivated or recombinant live vaccines.

The OIE takes the position that while vaccination can help suppress symptoms and death, reduce the amount of viruses excreted, and increase resistance to infection, it cannot impede the propagation of viruses through infection and is insufficient on its own if one is aiming for eradication, and so should be combined with monitoring of vaccinated flocks, strict hygiene management, and extermination of all infected birds.

In Japan, vaccination is in principle not employed to control AI in poultry. However, in the event that outbreaks occur successively in several farms within a movement control zone and prompt extermination is difficult, MAFF will consider vaccination. For such emergency situations, Japan has accordingly procured a stock of imported vaccine and has also developed its own vaccine.

Countries such as Indonesia, Vietnam, China, and Italy which have suffered repeated outbreaks are using vaccination as a means of control, but they are thought to be still a long way from total eradication.

Agricultural countermeasures to new types of influenza

I said that in terms of virulence in chickens, AIV can be divided into low and high pathogenic types, but both have infected people, making AIV a zoonotic disease. The HPAI H5N1 subtype that has been pandemic among bird species in recent years warrants special mention, having claimed 166 human lives between 2004 and early February in 2007.

The WHO fears that the H5N1 subtype of AIV will mutate into a new pandemic influenza virus that can easily infect human beings. Fortunately, it has still not mutated into a form that can be transmitted easily from one human being to another, but it is not hard to imagine such an eventuality should it continue to repeatedly make the jump from birds to humans.

In the cases reported so far, that jump has been from poultry to humans, and continued infection among poultry is a major cause of the emergence of new pandemic influenza viruses. As such, the only way for the agricultural sector to combat the emergence of new pandemic influenza viruses is to eradicate H5N1 HPAI in poultry. HPAI spreads easily across national borders, making strong international cooperation essential to its eradication. Without such cooperation, I don't think that we can prevent the emergence of new pandemic influenza viruses.

Chapter 19

Wild Bird Migration and Behavior in Relation to AIV Infection

Yutaka Kanai

Opening comments

The strongly virulent H5N1 AIV that emerged in Asia in the autumn of 2003 spread in 2005-2006 to Europe and Africa. It also reappeared in 2006 and 2007 in Korea and Japan. Wild birds are suspected of being the vectors in many cases of transmission, and there are cases of wild birds too being infected. However, there have been outbreaks whose location and period do not necessarily match wild bird migration patterns.

I want to take a look here at what current knowledge can tell us about the relationship between wild birds and the spread of AIV infection.

The relationship between outbreaks and migration routes

There are many unknowns about what kind of birds become infected with AIV. Even if migratory birds are infected, it would be difficult for them to travel long distances if suffering from the effects of the virus. If symptoms do not emerge, they could carry the virus for the distance that they travel in the approximately 10 days that it takes for the virus to disappear from their bodies. It is for such a reason that many people suspect that migratory birds have played a role in the worldwide spread of AIV. However, since most bird migration is limited to certain fixed periods and flight paths, it would be a mistake to single out migratory birds as the sole cause of the spread of AIV without first considering their behavior patterns.

1) From Korea to Japan

The AIV outbreak discovered on January 11, 2007 in Kiyotake, Miyazaki Prefecture followed the same pattern as the 2004 outbreaks in Yamaguchi, Oita, and Kyoto Prefectures of occurring during ongoing outbreaks in Korea. In both 2003-2004 and 2006-2007, the Korean outbreaks occurred in late November and early December, continuing into January.

In Japan in 2003-2004, the Yamaguchi Prefecture outbreak occurred in late December, and the Oita and Kyoto prefecture outbreaks occurred in early February, with each outbreak thought to have occurred independently. As for 2007, there have been 4 outbreaks to date (February 13) — 3 in Miyazaki Prefecture, and 1 in Okayama Prefecture.

Both the 2003 and 2006 outbreaks in Korea started in rice growing regions on Korea's west coast. Certain outbreaks occurred in almost identical locations. The outbreaks were not limited just to chicken farms, but occurred also on duck and quail farms. These farms were located near expansive waterfowl habitats, and on December 21, 2006, the same virus was found in wild ducks.

As the provided map shows, outbreaks in Japan have been scattered across southwest Japan. Another difference with Korea is that the majority of outbreaks have occurred at some distance from broad expanses of paddy fields and waterfowl habitats. The 3 Miyazaki outbreaks were close to each other, but occurred in different kinds of environments — one an area of fields with a fair number of houses, another a mountain valley, and the third a terrace bordering a riverside stretch of paddy fields.

The route from the Korean Peninsula to southwest Japan is a major migration route for wintering birds and is a journey of just 1 day. However migration is at its height between October and early December, and most wintering birds have arrived in Japan by mid-December. Birds arriving after this period tend to do so if the Korean Peninsula is hit by severe cold, with snow and freezing making the finding of food difficult.

There was no direct point of contact in Japan between poultry and the wild ducks considered to be the most likely vectors of the virus. During the 2004 outbreaks, we collected feces of wild ducks in areas surrounding the affected farms and also captured land birds to

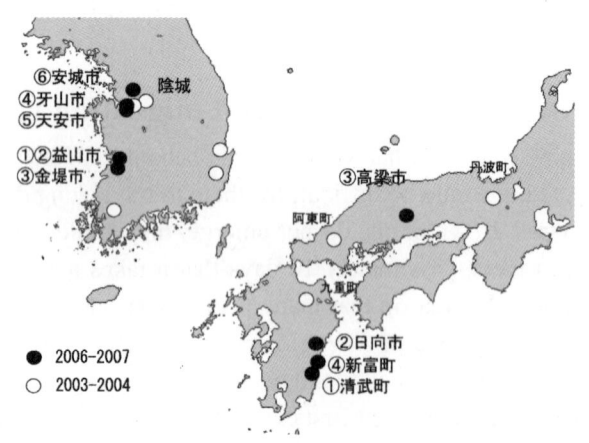

Fig. 1 Locations of strongly virulent AI outbreaks in Japan and Korea
The numbers on the map indicate the order in which outbreaks occurred in 2006-2007. The first AI outbreak in Korea in 2003 occurred in Eumseong, and spread to Cheonan. It was also in Cheonan that the H5N1 AIV was found on December 21, 2006 in the feces of wild ducks.

1,2	Iri	1	Kiyotake
3	Gimje	2	Hyuga
4	Asan	3	Takahashi
5	Cheonan	4	Shintomi
6	Anseong		Kokonoe
	Eumseong		Ato
			Tanba

investigate them for presence of the virus. However, the virus was not detected and there was no sign that infection had spread to land birds.

However, the possibility remains that a very small number of birds carry the virus. Flies and other insects can be found even in winter around the feces and feed boxes in poultry farms, attracting wagtails, thrushes, and other insectivorous birds. It is possible that active viruses accumulate in flies, so we need to look into the possibility that birds feeding on such flies could become infected with AIV.

2) The spread from Qinghai Lake to Europe

In the table 1, I show the relationship between the direction of spread of AIV and the period and migration since 2004. In both the 2005 and 2006 outbreaks at China's Qinghai Lake, bar-headed geese suffered infection and death, but there have been no reports suggestive of AIV from the species' wintering area in India. The key wintering areas of birds inhabiting Qinghai Lake are South Asian countries such as India and Bangladesh. In January and February of 2004, outbreaks occurred throughout China, including Lanzhou near Qinghai Lake and around Lhasa on the migration route. It would be natural to assume that the infected bar-headed geese became infected at Qinghai Lake or on the migration route near to the lake.

After the April 2005 outbreak at Qinghai Lake, infection spread to Europe, Africa, and India. There are those who argue that migratory birds are responsible for all of this spread, but such arguments are indefensible from the perspective of the behavior of birds. Before the outbreak in West Siberia, outbreaks had already occurred in the nearby extreme west of China. As stated above, by the winter of 2004, outbreaks had occurred throughout China.

Table. 1 Relationship between the spread of AIV infection and bird migration routes

Location of infection spread	Infection period	Relationship with bird migration
From Korea to Japan	Late December-early February	Main migration period is over, but because the distance is short, later migration is also feasible.
From Southeast Asia to Qinghai Lake	Mid-April	Not a major migration route
From Qinghai Lake to West Siberia	July	Not a major migration route, and period also differs.
From West Siberia to Black Sea coast and Middle East	September-October	Major migration route, and migration period overlaps infection period.
From Black Sea coast and Middle East to Nigeria	February	Migration route, but period differs.
From West Siberia to India	February	Major migration route, but period differs.
From Black Sea coast and Middle East to North Europe	February	Major migration route, and migration in February is possible.
From Southeast Asia to Russia, Alaska and Australia	-	Major migration routes, but no outbreaks along these routes

This suggests that AIV infection was spread through commerce.

The period when infection spread through West Siberia was the breeding season for birds, a time when they do not venture far from their nests. We should rather pay attention to the way AIV infection advanced along major routes of commerce from West China as the most likely reason behind the spread of infection.

Patterns of AIV spread that overlap with migratory routes and periods are the spread from West Siberia to Romania and other Black Sea coast locations, and from Romania and the Black Sea to North Europe. As for the spread of infection from migratory birds to poultry, it is thought that the shooting of wild birds that are then taken into homes to eat may be responsible.

In February 2005, an AI outbreak occurred in Nigeria. February is a time of the year not for southward migration, but rather for the start of northward migrations. BirdLife International has pointed out that Nigeria was a destination for the export of chicks from areas where outbreaks had occurred.

Cases of infection in wild birds

There are increasing reports of wild birds being infected with AIV, but almost no cases in which the infection route has been elucidated. The following are cases in which the circumstances surrounding infection are to a certain extent understood.

1) Kyoto Prefecture jungle crows (*Corvus macrorhynchos*)

It has been confirmed that jungle crows were secondarily infected during the February 2004 AIV outbreak in the town of Tanba in Kyoto Prefecture. Corpses of chickens that had died from AIV had been abandoned on the chicken manure dump of the affected poultry farm, and about 1,000 crows are thought to have fed on these corpses.

According to the survey carried out by Kyoto Prefecture and the Ministry of the Environment, there was no significant increase in crow deaths in the 6 roosts within a 30 km radius of the outbreak location. Of the 396 crow specimens found in Kyoto, Osaka, and Hyogo Prefectures from March to mid − April, only 9 were found to be infected, 7 of which were found on March 4 and 5 in Kyoto and Osaka Prefectures. No infected crows were found after April 5.

It appears likely that crows that fed on infected poultry were the only ones affected, and that infection did not spread directly between crows.

2) Thailand open−billed storks (*Anastomus oscitans*)

At Boraphet, a Thai swamp habitat and wildlife refuge, 496 open−billed storks died between January 18 and February 3, 2005. The outbreak occurred in an area of paddy fields about 30 km north of Bangkok that holds a nesting colony of over 5,000 open−billed storks.

The paddy fields where the storks feed are used also to raise free−range ducks for human consumption. It is thought that some individuals were infected by feeding on snails contaminated with the feces of infected ducks, and that these individuals spread the infection to others in the colony. The nests in the colony are so close together that they can become contaminated with the feces of individuals in surrounding nests, and the storks also gather in groups in ponds near the colony where it is thought that infection could spread easily via water. However, once infection among domestic waterfowl was controlled, infection among the open-billed storks abated.

Causes of the spread of infection to wild birds

1) Infection causes and wild birds

AIV infection of wild birds can take place in the following kinds of circumstances.

Intrusion into facilities for raising domestic fowl: Locations where feed has been scattered and flies and other insects abound tend to attract intrusion by wild birds seeking food. This needs to be prevented by the careful provision of feed and hygiene control. Intrusion can be prevented physically by covering windows and vents with netting or fencing with a mesh of under 3 cm.

Feces: If bird manure that is used as organic fertilizer on farmland contains AIV, infection is likely to spread over a broad area. In Southeast Asia, the scattering of chicken manure is a recommended method for boosting the productivity of fish farms, and there are those who argue that this is another cause of the spread of infection among wild birds. Ducks, herons, plovers, and curlews often visit such fish farms.

Refuse and remains: If refuse and corpses of infected fowl from poultry farms where an outbreak has occurred are left out in the open, crows, kites, vultures, and other wild birds will come to feed on such materials, and the possibility of infection is high. If the meat of infected birds is put on the market, infection in places other than poultry farms might occur.

Wild bird markets: Cases of infection have been discovered in hawk eagles and parrot species during customs inspections. These birds are thought to have become infected while in captivity. There is a high chance of infection spreading in wild bird markets where many different species from different locations are kept in close vicinity.

Other causes of spread of infection: Infected birds shot during hunting would be brought into the vicinity of human households, and hunting of birds also causes birds to scatter, promoting the spread of infection. Bird feeders, tables, and bird baths attract many individuals of different species, creating an environment for the spread of infection.

2) Research required for preventing infection

he following kinds of research are required to reduce the risks of infection from wild birds:

Virus surveys: Surveillance studies to check on the possession of AIV by wild birds are being carried out on an almost worldwide scale.

Tracking of migration routes and periods: Detailed information on migration routes and periods would enable the implementation of a more appropriate emergency response, including the enhancement of measures to prevent wild bird intrusion into poultry farms in areas along migration routes and hunting restrictions according to the time and location of an outbreak. Intensified banding surveys and satellite tracking are being carried in China, Russia, Japan, the USA, and other countries to gather more information on migration routes.

Consideration of measures to prevent infection among wild birds: Research on the types of wild birds living in the vicinity of poultry and their ecology and behavior would help to reveal points of contact between wild birds and poultry or other sources of infection.

Closing comments

Strongly virulent H5N1 AIV poses a threat to wild birds too, and while the culling of wild birds has been carried out on occasion, such measures adversely affect the wild bird population and ecosystem as a whole, and also carry the risk of exacerbating infection.

Controlling AIV requires a detailed understanding of the relationships between the virus, poultry, industry, and wild birds. We need to formulate countermeasures based on the recognition of AIV as a matter of ecosystem management on both a regional and a global scale.

Useful Internet links regarding wild birds and HPAIV

Wild Bird Society of Japan: http://www.wbsj.org/
Bird Life International:
   http://www.birdlife.org/action/science/species/avian_flu/index.html
Royal Society for the Protection of Birds (RSPB):
   http://www.rspb.org.uk/policy/avianinfluenza/index.asp

Chapter 20

Infection of Wild Birds and Current Status

Kumiko Yoneda

Highly pathogenic avian influenza (HPAI)

Japan has recently suffered its first outbreak of HPAI in three years. AI broke out suddenly in 2004, 79 years after its first appearance in Japan, and spread throughout Southeast Asia in roughly the same space of time, later spreading also to Russia and then Europe and Africa. A great deal of research has been carried out in the past 3 years that has revealed how the virus causing this disease has mutated.

A textbook available at the time of the 2004 pandemic has the following to say about AIV — that it is a virus carried asymptomatically mainly by wild ducks and other waterfowl, and exists in many different strains. In general the AIV harbored by waterfowl displays very weak infectivity with respect to poultry, and is also only weakly pathogenic. However, there are some viruses that, once they have succeeded in making the jump to chickens, mutate over several generations into strongly virulent strains. These strains are known as highly pathogenic avian influenza viruses (HPAIV), and are distinguished as such from other AIV strains in the formulation of poultry infection countermeasures. All HPAIVs to emerge so far have been of either the H5 or H7 type. There is only 1 case so far in which a large number of wild birds have died as a result of infection with AIV?the death in 1961 in South Africa of over 1,300 terns from the H5N3 HPAIV subtype.

As can be gathered from the fact that the vaccine for protecting against human influenza needs to be changed every year to keep in step with the latest strain, the influenza virus mutates very easily, and at the genetic level, no two identical viruses can be found even within the same epidemic. There are also many strains of AIV H5N1. GenBank, a database of gene sequences, shows 1,642 registered strains of AIV H5N1 isolated since 1959 as of July 16, 2007. However, only 11 had been isolated up to 1991, all of those from the UK and the USA. Of these 11, the 4 strains isolated from wild birds were nonpathogenic to chickens, and also produced no symptoms in the wild birds from which they were isolated. Owing to various factors such as advances in gene sequencing technology, increased surveillance, and the fact that researchers register strains on this gene database at their discretion, the number of strains registered does not necessarily reflect the reality at any particular point in time, but since 1997, the number of Asian strains, and particularly those from China, has risen dramatically. Several hundred strains have been registered each year since 2004 when outbreaks in Japan occurred. Strains isolated from wild birds (including captive individuals) number about 240. All of this suggests that some major change occurred around 1997.

AIV outbreaks in wild birds

In 1997, an outbreak of HPAIV of the H5N1 subtype killed a great many chickens and other domestic fowl in Hong Kong, and also infected humans. It was determined that this strain had mutated from a strain isolated from domestic geese in the Chinese province of Guangdong in 1996. The same strain further mutated into a different strain that caused the death of 150 waterfowl in two parks in Hong Kong in 2002. This event overturned the widely accepted belief at the time that AIV displays no pathogenicity in wild waterfowls and other wild birds. Both outbreaks happened in parks within the city, with one affecting mostly captive wild birds, as a result of which there is a detailed report on the progress of the outbreak and species that did or did not succumb to the disease. According to this report, waterfowl death rates were high for wild geese and Greater flamingos. Among the captive wild ducks, fatalities were high for New World species and red-crested pochards among Old World species. It was also determined that there were species in the same park ponds that suffered no deaths despite being infected, and other species that suffered neither deaths nor infection, suggesting that pathogenicity of HPAIV to wild geese and ducks differs according to species.

The next incident of mass death among wild waterfowls occurred at Qinghai Lake in inland China in May 2005. Within the space of 2 months, about 6,000 birds died, with bar-headed geese being the main victims, but with deaths also occurring among brown-headed gulls, great black-headed gulls, ruddy shelducks, and great cormorants. In August of the same year at two lakes near Mongolia's border with Russia, about 90 bar-headed geese, whooper swans and other waterfowl died from AIV. Repeated outbreaks occurred in the same areas in 2006, with bar-headed geese deaths in the main being reported in Qinghai Province in April and May, and bar-headed geese and wigeon deaths in Tibet in May. In June, swan, gull and geese deaths were reported at lakes near Mongolia's border with Russia.

Meanwhile in Europe, repeated deaths were reported mainly among mute swans between October 2005 and May 2006. Particularly in Germany, many AIV deaths were reported for swans, wild geese and ducks, raptors, and other types of wild bird. All of the virus strains causing death in wild bird species since the Qinghai Lake outbreak have proven to be very similar and traceable to the so-called Qinghai strain.

Susceptibility of wild birds to HPAIV H5N1 subtype

Differences in susceptibility to HPAIV among bird species differ also according to strain. Chickens, turkeys, and other members of the order Galliformes were very susceptible to the pre-2002 H5N1 subtype strains, but domestic ducks and geese showed only low susceptibility. However with the emergence of the HPAIV that spread throughout Southeast Asia in 2004, domestic ducks started to die from AIV infection, and as described above, mass deaths were reported also among wild geese, wild ducks, and gulls. There have also been reports?for example, the Kyoto Prefecture outbreak — of secondary infection of wild bird species during an outbreak among chickens and other poultry.

Pathogenicity with respect to chickens can now be ascertained from the sequence of amino acids in the virus genome, but this is not yet possible for wild geese and ducks. However, it seems reasonable to assume that if pathogenicity experiments were conducted on domestic geese and ducks, results could be used to judge whether the same virus strain would be pathogenic or not to wild geese and ducks. There are only a few reports of infection experiments conducted on wild bird species. This is because it is not only difficult to obtain healthy wild bird specimens, but it is also difficult to keep such specimens in isolator cages for prolonged periods. Reports of HPAIV outbreaks among free-ranging wild birds usually name the species that have suffered deaths, but rarely mention the species living in the same habitat that did not suffer infection or death. It is however possible to deduce the species likely to have been present by referring to habitat reports and other literature published prior to an outbreak. A look at about 130 species that were reported for HPAIV infection, for the circumstances of outbreaks, or for the results of pathogenicity experiments suggests that there are wild bird species with low susceptibility to the HPAIV H5N1 subtype.

There are those who argue that where wild ducks, the original AIV host, are concerned, AIV will evolve to become less and less pathogenic. Any species that can be infected with AIV without suffering any symptoms but shed the virus can become an effective vÿÿtor for spreading the virus.ÿÿe need to continue to monitor wild birds for possible infection and carry out further research on species difference in susceptibility in order to develop a clear and accurate picture of how wild birds figure in HPAIV infection.

Chapter 21

The Threat Posed by New Types of Influenza: AIV and its Impact on Humans

Nobuhiko Okabe

Influenza is a well-known human disease caused by infection with the influenza virus. There are three types of influenza virus — A, B, and C. Types B and C are limited to humans, but type A is found in many different subtypes (currently 144 have been identified), only 2 of which are currently capable of infecting humans. The other influenza type A subtypes infect birds and many other animals.

Type A human influenza viruses

Type A influenza viruses are genetically unstable, and the H3N2 and H1N1 subtypes that are currently transmitted between people continue to mutate very gradually. This kind of gradual change is known as antigenic drift. Once the virus mutates into something new, even people who have suffered influenza previously may once again be susceptible to infection by the new mutation.

Rather than remaining the same subtype for tens or hundreds of years, type A influenza virus has tended to change suddenly into another subtype every few years or decades. These changes are not small-scale changes within any one subtype, but rather changes on a scale that produces a new subtype. This is called antigenic shift, and could be thought of as something like a full model change in auto industry parlance, a totally new human influenza virus. Since it is new, nobody has any resistance to it, and so for a while widespread pandemics ensue. Spanish flu (H1N1) swept the world in 1918 as a huge pandemic, but then became established as ordinary influenza for the next 39 years, until replaced suddenly by Asian flu (H2N2) in 1957. This type reigned for 11 years before being ousted by Hong Kong flu (H3N2) in 1968, a new influenza virus that was subsequently joined in 1977 by Russian flu (H1N1). These two subtypes cause ordinary flu to this day while continuing to mutate very gradually.

Avian influenza virus

Birds are the leading example of animals that are susceptible to infection by type A influenza virus. The influenza viruses that infect birds are of the same type A that also infect humans, but their genetic structure differs slightly. Another difference is that AIV tends to replicate mainly in the intestines, which means that large quantities of virus are expelled with feces. Response differs according to type of bird. Wild ducks and other waterfowl can be infected by all 144 type A subtypes, but do not show symptoms. Because infected migratory ducks remain healthy as they fly from here to there throughout the world, excreting viruses along with their feces as they go, it is thought that ducks are the most likely cause of the spread of AIV. Some of the viruses expelled by ducks are thought to infect chickens, domestic ducks, and other poultry at watersides and feeding areas. In most cases such infections are by low pathogenic viruses (LPAIV) that produce only light or no symptoms at all in most chickens and ducks, but certain subtypes (H5 and H7 subtypes known as highly pathogenic AIV) are extremely pathogenic in chickens, with H5N1 in particular causing severe symptoms that rapidly kill almost all infected chickens.

Infection of humans by AIV

It was thought for a long time that AIV could never directly infect humans, but during the 1997 outbreak of AIV H5N1 in Hong Kong, the first ever case of human AIV infection occurred, with 18 people being infected, 6 of whom died. After the Hong Kong government exterminated the 1.5 million chickens thought to have been the source of infection, no further cases of human AIV infection occurred for a while, but then in 2003 in Holland, an outbreak of H7N7 AIV occurred among chickens that also produced symptoms (mostly conjunctivitis) in about 100 humans, and caused 1 human death. In 2005 there was an outbreak of AIV H5N2 in Ibaraki Prefecture, Japan, but no people fell ill on this occasion.

When it directly infects humans, AIV H5N1 produces very severe symptoms (60% death rate). However the virus does not readily infect humans, and almost all of the victims to date had close contact with birds that were suffering or had died from the virus. There was no spread of the infection in hospitals caring for AIV-infected patients or transmission from people to people in the vicinity of outbreaks, and nor was there any transmission through poultry food products. In short, in a country like Japan, for example, no one leading a normal life has any reason to fear catching AIV just because they happened to spot some birds flying past their windows.

The threat posed by the emergence of a new influenza virus

As mentioned earlier, Spanish flu (H1N1 subtype) continued to circulate for 39 years from 1918, and then Asian flu (H2N2) for 11 years from 1957. In 1968, Hong Kong flu (H3N2) appeared, joined in 1977 by Russian flu (H1N1), and they remain in circulation to this day, the former for almost 40 years, the latter for almost 30. Both types continue to mutate slightly, but they are still basically the same type A influenza viruses that they started out as.

If you look at the history of change of influenza up to now, it would not be in least surprising for the type A virus to undergo another full model change at any time to give rise to a new influenza virus, but no one can forecast with any scientific accuracy whether this event will take place next year or 5 years or more from now.

However, if a completely new influenza virus does appear, humans will not have any resistance to it, and there are consequently fears that it will cause a worldwide pandemic. While medicine too has advanced, the development of transport, increase and concentration of human population, changes in lifestyles, and so forth make this a very different world from the one in which previous pandemics occurred, and if and when a new virus appears, it is sure to spread at an unprecedented pace and become a huge pandemic. This is giving rise to serious concern over increases in human illness and the severe impact that a major pandemic would have on human society.

Theories regarding the emergence of a new influenza virus include: (1) a genetic mutation of AIV may result in a new subtype that can easily infect humans; and (2) AIV and the human influenza virus may simultaneously infect either pigs or humans (pigs are able to be infected by both human and avian influenza), and as a result of mixing of the genes of both viruses (gene exchange), a new subtype may be created. The occurrence of major outbreaks of AIV among birds and the concomitant rise in number of human infections, rare and coincidental though they might be, is thought to point to an increased risk of a new influenza virus emerging.

Currently H5N1, which is the cause of increasing outbreaks of AI among poultry, is seen as the most likely precursor of a new human influenza virus. The possibility of the emergence of such a virus through pigs remains, so we are continuing to monitor occurrences of influenza in pigs.

AIV and new influenza virus countermeasures and preparations

To regard the appearance of a new influenza virus as unlikely is an excessively optimistic view. At the same time it is unrealistic, given our current capabilities, to think that we can totally prevent such an eventuality. I think we have to accept it as a part of the natural order of things. However, just as we prepare for earthquakes and other disasters, we should of course take steps to curb the scale of any epidemic that occurs, and minimize damage to health and impact on society.

If we keep going on endlessly about the possibility of a new influenza virus appearing, we run the risk of being seen as Aesop's fabled "boy who cried wolf", but whatever preparations we make need not be limited just to countering a new influenza virus. We will be able to apply them equally to other new threats such as SARS, and so they are vital as countermeasures to all infectious diseases, whether known or as yet unknown.

Chapter 22

Highly Pathogenic AIV and Vaccination Measures

Tetsuo Nakayama

In 1997 in Hong Kong, 18 people were infected with AIV H5N1, and 6 of those died. It was determined that the virus had been transmitted from chickens, and the spread of the infection was prevented by the extermination of the chickens thought to have been its source. Later, H5N1 transmitted from wild ducks spread among domestic ducks and geese in China and other parts of Southeast Asia, and then came to infect poultry such as chickens and quails, and also wild birds. As a result, H5N1 spread to the Middle East and Europe and became a global issue. Since 2003, human infection with AIV has been reported in Vietnam and Thailand, and the number of human infections has risen to almost 270, with a death rate of over 50%. There are cases in which human-to-human transmission could be suspected, but because these cases involved members of the same family sharing the same environment, there is little strong support for the human-to-human infection argument. In Japan too, AIV outbreaks resulting in chicken deaths occurred at poultry farms in Kyoto, Yamaguchi, and Oita Prefectures in 2004, and occurred again in 2006 in Miyazaki and Okayama Prefectures, prompting fears of human infection. Anti-influenza drugs have been developed, but as a preventive measure, the development of vaccines is fundamental to combating influenza.

Influenza vaccine is manufactured using fertilized eggs containing 15 μg of HA protein respectively for A/H1, A/H3 and B strains. Because HPAIV kills fertilized eggs, conventional production methods are difficult. A phase I study has been completed in which pharmaceutical companies have each developed prototype HPAIV vaccines from vaccine seeds of recombinant viruses created by using genetic engineering techniques to modify HA protein genes related to pathogenicity and reduce their virulence, and then combining them with genes other than HA and NA replicated in eggs. Each company conducted clinical trials in which 6 groups of 20 (subcutaneous and intramuscular; 1.7-, 5- and 15-μg HA dose groups respectively) were vaccinated twice with whole inactivated viruses to which aluminum adjuvant had been added. The subcutaneous groups showed strong local reactions, but overall reactions were much the same as for existing vaccines. Immunogenicity was confirmed for doses of 5 μg and above. Based on the phase I study results, the three front-running companies conducted phase II and phase III studies. The results of clinical trials using 4 groups (subcutaneous and intramuscular; 5- and 15- μg dose groups respectively) and a total of 900 subjects, cleared the EU's immunogenicity criteria for inactivated influenza vaccines.

The prototype vaccine HA protein was of a strain isolated in Vietnam in 2004 (clade 1), but the influenza virus mutates rapidly, and the strain currently circulating is a clade 2 strain which is further subdivided into subclades 1, 2, and 3 as its antigenicity changes. We are currently manufacturing a store of pandemic prototypes using a strain derived from clade 2.

I would like to explain the limitations of current influenza vaccines and the global status of development of vaccines against new influenza viruses.

Part IV
Effect of Cadmium and Arsenic on Agriculture, the Environment and Health

23	**A Message from the Symposium Organizer** ················· 105 *Tadayoshi Shiba*	
24	**Biogeochemical Cycle of Heavy Metals : Cadmium and Arsenic** ············ 107 *Katsu Minami*	
25	**Risk of Heavy Metal Contamination of Farmland Soil and Countermeasures** ·································· 112 *Shinichi Ono*	
26	**Accumulation of Heavy Metals in Plants and Intake by Humans** ············ 116 *Tadakastu Yoneyama*	
27	**Status of the Codex Alimentarius Commission and Japan's Approach** ········ 120 *Masahiro Segawa*	
28	**Assessment of the Biological Effects of Cadmium Intake : Attempt at Estimation of Tolerable Intake** ····················· 126 *Hisayoshi Ohta*	
29	**The Battle of the Codex: Standards and Food Safety ; The Case of Cadmium, Chloropropanol and Formaldehyde** ············ 130 *Fujio Kayama*	

30 **Heavy Metal Problems from the Perspective of Clinical Ecology** ············ 133
Kou Sakabe

Chapter 23

A Message from the Symposium Organizer

Tadayoshi Shiba

It is my honor to welcome you to Kitasato University's 4th Agromedicine Symposium.

Dr. Shibasaburo Kitasato has made immeasurable contributions to the development of modern medicine and health services in Japan. Dr. Kitasato began his "*Idoron* (Medical Ethics)" (1878), which he wrote when he was just 25 years old, by expressing his belief in medical ethics:

> "Ancient people said medicine is the healing art. They also said it is virtue that a great doctor heals the nation. Medicine in its true form is to help people maintain their health, work with peace of mind, and develop an affluent and strong nation. If one knows no good regimen, he cannot maintain his body in good health. Without health, living is hardly meaningful...fundamental to medical ethics is for practitioners to help people understand how important health is so that they will take care of hygiene and prevent illnesses before they strike."

Dr. Hisayuki Omodaka, who lectured at the medical faculty of Kitasato University in its early years, was deeply involved in the "Principles of Medicine" course that is presently offered at the faculty. In his book "*Igaku Gairon Towa* (An Outline of Medicine)" (Seishin Shobo, 1987), Dr. Omodaka roughly says this:

What does medicine study? Not the philosophy of life. Not the ethics of medicine (although it is part of the outline of medicine). Not just medical ethics. Medicine considers not only the phenomena of physical life but also spiritual phenomena. Medicine cannot be just one of the natural sciences. It must be a social science as well. Medicine is knowledge of as well as the art of healing illnesses. Medicine studies not only the ways to treat or prevent illnesses, but also the ways to maintain health. Medicine is, however, not only a discipline for maintaining health. It must be a discipline for willingly improving health.

These books by Drs. Kitasato and Omodaka tell us that medicine is a discipline that should encompass the treatment and prevention of illnesses, the maintenance and improvement of health, as well as the solution of spiritual aspects. In order to fulfill these mandates, it is important for us to maintain wholesomeness and safety of food (agriculture) and the environment that are the basis for the life. These books tell us that people cannot be healthy without wholesome food and a sound environment. Our predecessors have already spoken of the need for a science in which agriculture and medicine collaborate through environmental efforts.

From these perspectives, we at Kitasato University, which aims to be at the frontier of life science, promote the close collaboration of agriculture, environment and medicine. We direct our efforts in education and research to the issues pointed out by our predecessors, as well as new issues, such as infectious disease, food safety and global warming, that modern society faces today. The Kitasato University Agriculture and Medicine Symposium is part of our efforts.

The 4th symposium focuses on the behavior of arsenic and cadmium from the perspectives of food safety, biogeochemistry, soil, plants, clinical environmental medicine and law. We hope the symposium will help in the development of the collaborative science of agriculture and medicine.

During the symposium, we hope you will engage in meaningful and practical discussions that will lead to new ideas and suggestions for dealing with health issues arising from issues relating to food and the environment. We would like to extend our sincere appreciation to those of you who have agreed to speak at the symposium.

Chapter 24

Biogeochemical Cycle of Heavy Metals : Cadmium and Arsenic

Katsu Minami

"All substances are poisons: there is none which is not a poison. The right dose differentiates a poison and a remedy."

Paracelsus (1493–1541)

Introduction

The modern civilization in which we live today depends on a tremendous quantity of heavy metals to exist. History shows a close correlation between the evolution of humankind and the consumption of heavy metals.

Humans began using copper in 6000 BC approximately, lead in 5000 BC and zinc and mercury in 500 BC. These events are evidenced by analyses of heavy metals contained in sediments, cores of polar ice and peat. History has proven the impacts of heavy metals on environment. It has been also verified that the Roman Empire used a considerable quantity of lead.

With the start of the Industrial Revolution in the 19th century, heavy metals became increasingly essential to modern society. As a result, the types and quantities of heavy metals extracted from the earth increased, inevitably leading to exponential increases in these metals being dispersed into the soil, vegetation, ocean and atmosphere. This began to disrupt the biogeochemical cycling of heavy metals.

What does the disruption of the biogeochemical cycle of heavy metals mean? Heavy metals that have been going through steady cycles would begin to impose an excess load on the atmosphere, soil and ocean. These excess heavy metals loaded into soil are absorbed by crops. The heavy metals dispersed in oceans are taken up by the fish and shellfish that live in the waters.

As a consequence, humans and animals that eat these crops, fish and shellfish begin to accumulate higher–than–normal amounts of heavy metals in their bodies. These heavy metals will pass down through future generations of humans and animals. The heavy metals will accumulate in humans through the food chain. It is an accumulation that will span generations. Heavy metal contamination is a problem that transcends time and space.

The purposes of agriculture and agronomy are to supply people with safe and sufficient food as well as biological resources for things such as clothing. In order to ensure these purposes are met, we must preserve the environment. The purposes of healthcare and medicine are to save people from diseases and to protect their health. In order to ensure these purposes are met, we must preserve the environment just as much as agriculture and agronomy are expected to. We cannot ensure production and health if we ignore the environment.

Many factors impede the production of food and biological resources, contribute to illnesses or adversely affect our health. One of the key factors is the contamination of the environment by toxic metals.

Some of these toxic metals, such as cadmium, may be toxic to animals, including humans, even at concentration levels that do not inhibit the growth of plants. As a consequence, the contamination of the environment by toxic metals is an issue that agriculture and agronomy, as well as healthcare and medicine, cannot sidestep.

Although localized, we have been unfortunate enough to experience this problem in illnesses such as the itai-itai disease caused by cadmium poisoning and the Minamata disease caused by mercury poisoning. Events such as these have the potential to occur everywhere on earth in the future.

The Codex Alimentarius Commission created jointly by FAO and WHO has already established international standards for food and legislated regulations of cadmium and other heavy metals in food.

It is essential for all life existing on earth for heavy metal concentrations to be limited to appropriate levels. This is particularly true for food that animals and humans consume. Heavy metals dispersed from the earth's crust into the natural world in the soil, ocean and rivers will ultimately be accumulated in human bodies through plants, fishery products and animals. Accordingly, we need collaborative studies among agriculture, environment and medicine to find solutions to these heavy metal problems.

This article will survey the behavior of cadmium and arsenic from the perspectives of biogeochemistry, soil, plant, clinical environmental medicine and law in the hope of assisting the collaborative science of agriculture and medicine.

Development of civilization and the dispersion of heavy metals

Many fields of studies are increasingly casting a spotlight on heavy metals as environmental problems become more global. To put it in a few words without fear of misunderstanding, humans have been mining large amounts of metals from the earth's crust and dispersing then over the earth's surface as their civilizations developed. Demand for metals accelerated at an unprecedented pace particularly during the Industrial Revolution. Until then, heavy metals had remained deep in the ground since time immemorial, quietly asleep.

A look at history tells us that humankind owes its development to heavy metals to a great extent. Modern civilizations could not have been established without large amounts of heavy metals. Analyses of heavy metals contained in sediments, cores from polar ice sheets and peat reveal the extent of the impact that heavy metals have had on the environment.

The Roman Empire needed massive amounts of heavy metals for its citizens to maintain their comfortable and lavish life. The Romans consumed 80,000 to 100,000 tons of lead, 15,000 tons of copper, 10,000 tons of zinc, and more than 2 tons of mercury annually. Tin was also in high demand. Mine business was small at that time but their smelting operations processed a large amount of ores in uncontrolled open systems, dispersing considerable amounts of trace metals into the atmosphere. This led to increasing types and amounts heavy metals that were mined from the earth's crust, inevitably resulting in increased dispersion into the soil, vegetation, oceans and atmosphere.

As a matter of course, the growth of world population and increases in heavy metal consumption that accompanied the growth resulted in dispersal of heavy metals into the natural world, creating a variety of ecological problems. Many of the heavy metals contained in soil, water and living organisms are essential for a healthy life, even though excessive concentrations have toxic effects on living systems. It is, therefore, important for us to know the appropriate concentration levels of heavy metals in living organisms that exist in nature, as well as in food that animals and humans consume. Heavy metals that were dispersed in nature will be ultimately accumulated in the human body through soil, vegetation and animals. Accordingly, it is also important to accumulate knowledge of the behavior of these heavy metals in soil. Indeed, this is an issue that requires collaborative studies among agriculture, environment and medicine.

Comparison of past and present distributions of heavy metals

Here are some specific numbers. A study by Hong et al. (1994) found that the ice sheet cores deposited in northwestern Greenland between 500 BC and 300 AD contained lead at a level that was 4 times higher than the background. This means that the contamination by lead dispersed from Roman mines and smelters spread over the Northern Hemisphere.

The lead content decreased to an earlier level (0.5 pg/g) after the fall of the Roman Empire, but began to increase again with the mining renaissance in Europe, reaching 10 pg/g in the 1770s, and 50 pg/g in the 1990s. The lead content in the arctic snow began to decrease in the 1970s, due perhaps to a switch to unleaded gasoline in North America and Europe.

Lead contamination of the atmosphere is not limited to the Northern Hemisphere. Woff and Suttie (1994) reported that the average accumulation of lead in the arctic snow during the 1920s (2.5 pg/g) was 5 times higher than the background level (less than 0.5 pg/g). The lead content is lower in Antarctica because the Southern Hemisphere generates less lead.

Studies of other types of sediments also revealed that lead contamination occurred on a global scale in ancient times. An analysis of sediments in various lakes in Sweden indicates that there was a peak in the build-up of lead around 2000 BC. The build-up gradually increased around 1000 BC, and reached a level that was 10 to 30 times higher than the background in the early days of the industrial revolution. The build-up of lead accelerated in the 19th century, and peaked in the 1970s (Renberg et al., 1994).

The records regarding ombrogenic bogs in Etang de la Gruère in Switzerland reveal that the lead build-up at its peak in 2000 BC was at the same level as it is in recent sediments. Similar peaks of build-up of lead in Roman times have been reported for European peat bogs, such as the Gordano Valley near Bristol and the Featherbed Moss in Derbyshire in England.

The world is being contaminated by a variety of metals. Indeed, it is a difficult time for us to fulfill our ethical objective of preserving a healthy environment for future generations. The following section will discuss cadmium and arsenic, which have had a large impact on human health. These substances have been investigated as the causes of environmental diseases, and reviewed by the Codex Alimentarius Commission as elements in food.

Cadmium and arsenic

There are four diseases in Japan that are officially recognized by the Ministry of the Environment of Japan as pollution-related diseases. They are itai-itai disease caused by cadmium (187 designated victims), chronic arsenic poisoning (188 victims), Minamata disease caused by organic mercury (2,995 victims) and respiratory diseases caused by air pollution (53,502 victims).

Of various pollution-related diseases addressed by the 4th Kitasato University symposium, cadmium and arsenic are the most discussed substances. They have been discussed at numerous symposia and investigated by numerous studies. There are a large number of reviews as well. Yet few attempts have been made at integrating the knowledge of biogeochemistry, agronomy, soil science, environmental science, clinical environmental medicine and law in any comprehensive form.

I will present to the symposium how cadmium and arsenic are linked through biogeochemistry to agronomy and soil science. The following section is a brief look at examples of arsenic and cadmium contamination around the world and in Japan.

Arsenic and cadmium contamination in the world and Japan

1) Regions with arsenic-contaminated aquifers, mines and geothermal water

Arsenic contamination of aquifers: United States (Western states), Mexico (North central, Lagunera), Chile (Antofagasta), Argentina (Chaco-Pampean Plain), Hungary-Romania (Great Hungarian Plain), Nepal (Terai area), China (Shanxi, Gui Zhou Province, Inner Mongolia, Shanxi, Xinjiang-Uygur Autonomous Region), Bangladesh (West Bengal), India (West Bengal), Vietnam (Red River delta), Cambodia (Mekong River), Myanmar (Ayeyarwady River), Pakistan (Indus River), etc.

Arsenic contamination caused by mining: Alaska (Fairbanks), Canada (British Columbia), seven regions in the United States (Coeur d'Alene, Clark River, Lake Owen, Wisconsin, Halifax County, Badger, Don Pedro), Mexico (Zimapan Valley), Brazil (Minas Gerais), Ghana (Asanti), Zimbabwe, England (Southwest), Poland (Southwest), Austria (Styria), Greece (Lavrion), Korea (Gubong), Thailand (Romphibun), Indonesia (Sarawak), etc.

<u>Arsenic contamination of geothermal water</u>: Dominica, El Salvador, USA (Alaska, western states), Chile (Antofagasta), Argentina (Northwest), France (Massif Centrale), New Zealand (Wairakei), Russia (Kamchatka), Japan (Miyazaki, Shimane), etc.

2) Areas designated for remediation projects for cadmium‐contaminated soil in agricultural lands in Japan

As of March 2006, there are 60 areas in Japan that are designated for remediation projects for cadmium‐contaminated soil in agricultural lands. The areas total 6,228 hectares. The concentration of cadmium in brown rice in these designated areas exceeds 1.0 mg/kg. The areas are situated in 22 prefectures, from Akita in the north to Kumamoto in the south. To date, remedial projects have been completed for 90% of the designated areas covering 5,618 hectares.

3) Areas designated for remediation project for arsenic‐contaminated soil in agricultural lands in Japan

There were 14 areas (totaling 391 hectares) across Japan where the detected concentrations of arsenic exceeded the standard prescribed by the Soil Contamination Control Act. Seven areas, totaling 164 hectares, were designated for remediation projects to clean up contaminated soil in agricultural lands as of March 2006. The status of these projects are described below. The arsenic concentrations in the soil of the designated areas exceeded 15 mg/kg.

Areas designated for soil remediation projects: Township of Kawauchi, Shimokita‐gun, Aomori Prefecture (13.5 ha; site delisted), city of Ota, Shimane Prefecture (7.3 ha; site delisted), city of Masuda, Shimane Prefecture (27.3 ha; site delisted), township of Tsuwano in Kanoashi — gun, Shimane Prefecture (66.1 ha; project completed), township of Ato, Mine‐gun, Yamaguchi Prefecture (8.4 ha; site delisted), township of Ogata, Ono‐gun, Oita Prefecture (27.7 ha; site delisted) and township of Takachiho, Nishiusuki‐gun, Miyazaki Prefecture (13.5 ha; project completed).

Of these designated areas, the prefectures of Shimane and Miyazaki have 21 and 167 persons, respectively, who have been designated as the victims of pollution‐related chronic arsenic poisoning.

Chapter 25

Risk of Heavy Metal Contamination of Farmland Soil and Countermeasures

Shinichi Ono

Introduction

From the discovery of mines early in the 8th century, metal mining in Japan grew into an industry from the late Middle Age to the early Modern Age. The Edo Period saw mine developments flourish. It was not until after the Meiji Restoration when the Japanese mines began to take the form of modern corporations. Major mines expanded under the direct control of the government during this period. Demand for copper and zinc as raw materials for weapons increased considerably, and so did the output of ores from mines. The demand for metals rose again after the Second World War because of the Korean War, and then accelerated in response to the high economic growth that began in the 1960s. Large volumes of metal ores were mined from the earth and refined in order to satisfy the increasing demands of society. Ore imports also began to increase to make up the shortfall in domestic production. Through these periods, various heavy metals were discharged into the environment in Japan, resulting in wide-spread contamination of the soil by cadmium and other heavy metals. Such contamination is a negative legacy of past human activities, which is demanding the expenditure of considerable funds and labor today for mitigation.

Heavy metal contamination of soil in agricultural lands

Although there is no accurate definition of what constitutes a heavy metal, a generally accepted collective term describes it to be a metal with a specific gravity above 4 to 5. The heavy metals that contaminate soil include cadmium (Cd), copper (Cu), arsenic (As), zinc (Zn), lead (Pb), mercury (Hg), antimony (Sb) and chromium (Cr).

Most of the contamination of soil by heavy metals originates in the contamination of water or air. Once the soil is contaminated, removal of the heavy metals is not an easy task.

The Agricultural Land Soil Pollution Prevention Act of Japan, enacted in 1970, designated copper, cadmium and arsenic as special toxic substances. Major incidents involving these three metals in the past, which caused soil contamination, are as follows:

The so-called Ashio Mine Poisoning Case, in which paddy fields and the croplands of the Watarase River basin were contaminated by copper discharged from the Ashio Copper Mine. This case, occurring in the mid-Meiji Period, is regarded as the beginning of pollution problems in Japan. In another case, cadmium discharged from the Kamioka Mines at the upstream Jinzu River was found to be the cause of the itai-itai disease that emerged among the residents of the river basin. Later, many more cases of cadmium contamination of the soil in agricultural lands in river basin areas with mines upstream emerged across the country. Flue gas from zinc and copper smelters contaminated the soil in neighboring farms with cadmium. Arsenic discharged from some of the mines in the Kyushu and the San'in Regions contaminated the soil in farmlands around the mines and impacted the growth of rice. Contaminated well waters affected the health of the residents.

The Ministry of the Environment (the former Environment Agency) continues to conduct surveys of the concentration of these three elements in farmland soil. Remedial measures, such as soil dressing, are being implemented through special soil remediation projects. Cadmium is the heavy metal that has caused the most serious contamination problems in recent years. Abatement programs are urgently needed.

The state of cadmium contamination and remediation

1) World trends in cadmium-related risk management

Although cadmium is not an essential element for plants, some plants take up cadmium through their roots and transport it to their edible parts. When the link between itai-itai disease and cadmium contamination became apparent in 1968, the then Ministry of Welfare and Health (now the Ministry of Health, Labor and Welfare; MHLW) revised the specifications and standards for food and food additives in 1970 to limit cadmium concentration in unpolished rice to less than 1.0 mg kg^{-1} under the Food Sanitation Act. At the same time, the Food Agency, as it was then known, set policy that the government would not purchase unpolished rice with cadmium concentrations exceeding 1.0 mg kg^{-1}. The policy also stipulated that the use of unpolished rice with a cadmium content of less than 1.0 mg kg^{-1}, which was deemed fit for purchase by the government, would be limited to non-food processing (e.g. industrial glue) if the cadmium concentration was more than 0.4 mg kg^{-1}. The Agricultural Land Soil Pollution Prevention Act came into effect in 1971; under this act, remediation projects using soil dressing were carried out for paddy fields that produced unpolished rice with a cadmium concentration of more than 1.0 mg kg^{-1}.

According to a survey by the Ministry of the Environment, the total area of farmlands designated as cadmium-contaminated areas across the country exceeded 6,000 hectares. Remediation has been completed on more than 80% of these lands to date, with remediation efforts continuing in the remaining areas.

Cadmium contamination of foods became a global issue in the 1960s. In 1998, the Codex Alimentarius Commission (a joint food standards commission of FAO and WHO; the Codex) developed a preliminary draft proposal to regulate cadmium concentrations in agricultural products.

The Codex continued to review its proposal on an annual basis, and adopted the proposed standards for wheat, potatoes, legumes (except soy beans) and vegetables at its general meeting in 2005. The 2006 general meeting of the Codex set the standard for polished rice at 0.4 mg kg^{-1}.

2) Soil remediation technologies

(a) Soil dressing

Soil dressing is an effective engineering technique, which aims to separate the roots of crops from the contaminated soil by introducing non-contaminated soil to the land. It is, however, costly and difficult to secure necessary material (i.e. non-contaminated soil).

(b) Water management and use of materials to inhibit cadmium absorption by rice plants

When the soil is in a reduced condition under flooded paddy fields, cadmium loses much of its water solubility as it binds to sulfur and becomes cadmium sulfate (CdS); when the soil is in an oxidized condition after water is taken out, cadmium is ionized as cadmium sulfate ($CdSO_4$), and dissolves in water. In other words, maintaining the paddy fields flooded to the extent possible to prevent the soil from drying could prevent cadmium from dissolving into the water. As a result, rice plants will take up less cadmium.

(c) Phytoremediation

Some plants have been known to take up cadmium effectively from the soil. For example, the tall golden rods of the chrysanthemum family and field pennycress of the mustard family are said to take up considerable amounts of cadmium. Sorghum of the Poaceae family and kenaf of the hibiscus family have been cited as other examples in recent years. Cadmium in the soil may be removed by growing these plants in contaminated agricultural land and allowing them to take up the chemical. This technique is called "phytoremediation", and is receiving considerable attention as an environmentally-friendly soil remediation technology.

The National Institute of Agro-Environmental Sciences (NIAES) recently found some varieties of rice in indica and japonica-indica hybrid cultivar groups that take up large amounts of cadmium. NIAES is currently investigating its availability for use in phytoremediation. The plants cultivated as part of a phytoremediation project will be harvested, transported away from farmlands and incinerated. The cadmium will be recovered from the ash.

(d) Chemical cleansing of soil

There is a method for the elimination of cadmium from soil by washing the contaminated soil with materials such as ferric chloride and water. After the cleansing, cadmium that was released to the surface of paddy water is collected by pump, filtered and recovered using a chelating resin. Rice will grow normally in the cleansed paddy fields and the cadmium concentration in unpolished rice will decline.

Contamination by lead, arsenic and other metals

1) Contamination by lead

Studies of isotope ratios of lead revealed that lead contained in exhaust gas emitted by motor vehicles burning leaded gasoline until the 1970s was a major cause of contamination of the soil along the roads. The uptake of lead by crops is generally very small. Crops grown in soil with a high lead concentration tend to have a relatively higher lead content, with the most concentrated lead in the root. Lead rarely migrates above the ground, especially into the fruits. Deposition from the atmosphere rather than uptake from the soil was blamed for the lead contamination of leafy vegetables.

2) Contamination by arsenic

In some areas in Japan detected arsenic content exceeds the standard prescribed by the Soil Pollution Prevention Act total 14 (391 ha). Of these areas, seven (164 ha total) were designated as areas requiring remediation under the law. Although arsenic inhibits the growth of rice in arsenic-contaminated paddy fields due to its toxicity, it rarely migrates to unpolished rice, except for some organic arsenic that appears to be taken up rice plants. Recent studies have detected diphenylarsine (DPAA) and phenylmethylarsine (PMAA) in unpolished rice grown in organic-arsenic contaminated paddy fields. Yet, it is scarcely known how rice plants take up organic arsenic.

3) Contamination by other heavy metals

Other heavy metals that are of concern to humans as soil contaminants are zinc, copper, mercury, antimony and chromium. Zinc and copper are essential elements for both plants and animals, and pose little threat to crops unless their concentrations reach a very high level. As plants take up little mercury, antimony and chromium, these metals in soil of agriculture lands are of little concern with respect to crop contamination.

Chapter 26

Accumulation of Heavy Metals in Plants and Intake by Humans

Tadakastu Yoneyama

Introduction

Man needs nutrients to grow and maintain a healthy body. We obtain essential nutrients mainly from food. Yet, we need a large amount of water, and therefore, we ingest components that are dissolved in the water through the mouth in the same way as food. Heavy metals enter the human body as minerals contained in food, and as heavy metals dissolved in water (Fig. 1).

Some of the heavy metals that enter the human body, such as iron (Fe), zinc (Zn), copper (Cu), trace amounts of chromium (Cr) and selenium (Se), are essential nutrient minerals, while others, such as cadmium (Cd) and arsenic (As), are toxic to humans. Heavy metals enter the human body primarily through food and water. Those entering through water are in the form of free ions dissolved in the water, while those entering through foods are bound to components of the food. As illustrated by Figure 1, most of our food, except fish and shellfish, is produced in agricultural lands. Food grown in agricultural lands directly enters our mouth. Feed consumed by livestock also enters our mouth indirectly as livestock productsFoods and animal feeds are grown in the soil of agricultural lands. The soil contains heavy metals that exist in nature as well as those that enter the soil through irrigation and fertilizers. All are taken up by crops.

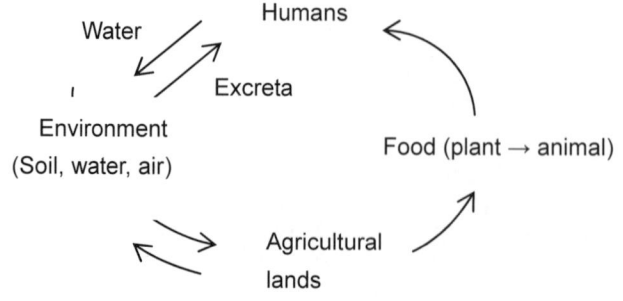

Fig. 1 Accumulation of heavy metals in food and intake by humans

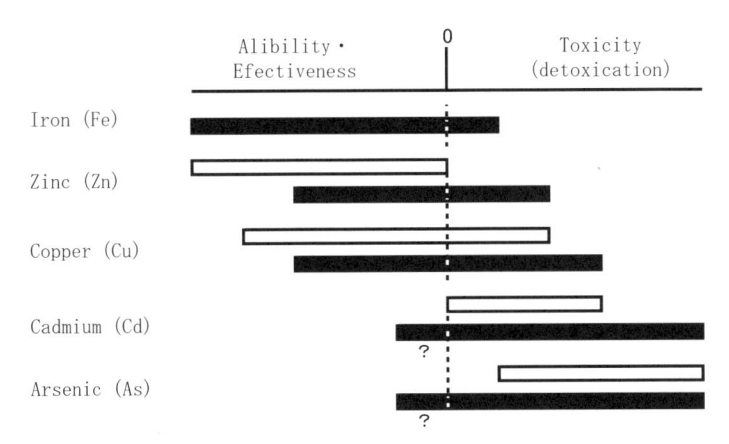

Fig. 2 Nutrient and toxicity of heavy metals to plants and humans
Plants ☐   Humans ■

Living plants take up essential trace elements such as Fe, Zn and Cu, for their growth and maintenance of their functions. The plants also take up non-essential heavy metals, such as cadmium and arsenic, from the environment (i.e. soil), (Fig. 2). I am a specialist in plant nutritional science; I study mechanisms for the absorption of heavy metals by plants, as well as the functions and toxicity of these heavy metals in the plant's body. In this article, I will focus on cadmium and arsenic, which are toxic to both plants and humans.

Accumulation of cadmium in plants

Japanese people ingest about one-half of their cadmium from rice, with the remainder coming from other foods. Grains and vegetables take up dicationic cadmium dissolved in the soil solution. The cationic heavy metals that are taken up by plants include Fe^{2+}, Zn^{2+}, Cu^{2+} and Mn^{2+}. A cation transporter that acts on uptake has been identified for each of these heavy metals, except for one specific to the uptake of Cd^{2+}. The above-described dicationic transporter seems to take up Cd^{2+} incidentally

Cd^{2+} that was taken up by the root binds with anions, such as organic acids, at an apoblast site, and is transported to the xylem situated at the center of the root. Cd^{2+}, which enters through a symplast site moves more slowly because it binds with glutathione, phytokeratin or metallothionein. Rice plants accumulated 90% of the absorbed cadmium in the roots. Cadmium that is transported up the xylem reaches foliage and fruits.

Approximately 1% of the cadmium that is taken up into rice plants is distributed to the fruit (i.e. rice grain). How does cadmium reach rice grains? Is it possible to suppress this migration into rice grains? These questions present interesting tasks for us to tackle. Cadmium has been thought to migrate directly to rice grains through the xylem, or is transported through the xylem to leaves and then to the rice grains through the phloem. Recent studies by the author and his team found that cadmium is transported through the connected phloem, since rice grains are not connected to the xylem, and cadmium is not present in the form of free ions but was bound to protein in the weakly alkaline fluid of the phloem. Since most of the cadmium that is transported from the root to the above-ground part through xylem is ultimately transferred to rice grains thorugh the phloem, my team estimated that cadmium in the xylem crosssed over to the phloem at a node. This xylem-to-phloem transport is found for other nutritional elements. When Cd^{2+} reaches the leaves, it is believed to bind to phytokeratin (PC) or metallothionein, and, in rice grains, to glutelin, a protein.

The suppression of the uptake of cadmium by the root and the migration to rice grains through the phloem may reduce the accumulation of cadmium in rice grains. Accumulation in leaf vegetables, on the other hand, may be reduced by limiting the availability of cadmium for the roots to take up and suppressing the transport of cadmium from the root through the xylem to the leaves.

Although certain plants that accumulate cadmium in a high concentration (hyperaccumulators) have been found in the areas of high cadmium concentrations, these plants are small in size. They can survive on cadmium-contaminated lands because they posses a system with which to detoxify the absorbed cadmium. It may be possible to clean up cadmium-contaminated soil by adding such a detoxification system to larger plants (phytoremediation).

Accumulation of arsenic by plants

People living in areas contaminated by arsenic take up arsenic mainly through drinking water, and a very little from food. Arsenic in fertilizers applied to the soil is absorbed into food and migrates to humans. In Bangladesh where the groundwater is contaminated by arsenic, drinking water as well as rice, wheat and vegetables grown with the irrigation of contaminated groundwater are reported to be contaminated by arsenic. Arsenic taken up by plants is anionic arsenic bound to oxygen as an arsenate (H_3AsO_4, H_2AsO^{4-}, $HAsO_4^{2-}$). They are As(V) that resembles phosphates and arsenites (As(III)) in the form of the neutral $As(OH)^3$. Anionic minerals taken up by plants include bromates ($B(OH)^{4-}$), molybdates (MoO^{4-}), as well as neutral mineral $B(OH)^3$ and silicates specifically taken up by rice plant ($Si(OH)^4$), which are respectively taken by anion transporters or aquaporin. Arsenates are taken by a phosphate transporter in the cell membranes of plants. Once inside the plant, As(V) is reduced to As(III) by arsenate reductase, and binds to phytokeratin containing a sulfhydryl (SH) group or glutathione (AS(III)-thiol aggregate) to be detoxified. Since the arsenates remaining as As(V) are analogous to phosphorus, they inhibit the production of adenosine triphosphate (ATP).

Reduction of the accumulation of arsenic in rice grains and vegetables can be achieved by suppressing the uptake by the roots of arsenic and reducing the transport rate of the absorbed arsenic to the parts above ground.

A recent study reported the discovery of a pteridophyte (*Pteris cretica*) that accumulates arsenic at a high concentration (Ma et al., 2001).

Human intake of cadmium and arsenic

Cadmium in food is bound to organic matter. It migrates to rice grains by binding to a metallothionein-like protein with a high cystine content. In albumen, cadmium is bound to protein. Kitagishi et al. reported a bond with gluterin (1976). Kitagishi is a pioneer in Japan in analysis of heavy metals in crops. Cadmium binds to citric acid in roots and accumulates in leaves, where it binds to phytokeratin, a compound that contains SH, and then accumulates in vacuoles within cells. We ingest cadmium bound to proteins or phytokeratin, which is likely absorbed as cadmium ions by the intestines. The absorption rate of cadmium in food is thought to be 2 to 8%, which is about one-tenth of the absorption rates of iron (60%) and zinc (75%) in food. A recent study by Horiguchi et al. (2004) reported that the cadmium absorption rate was correlated to age. The rates were 44% for those aged 20 to 30 years old, 1% for 40 to 59 years old and -5.9% for 60 to 79 years old, with an average of 6.5% for all age groups.

Arsenic exists in food as an aggregate with phytokeratin and inorganic arsenates, and as arsenate (As(V)) and arsenites (As(III)) in drinking water. Humans take in arsenic in these forms. It has been reported that pigs take up about 80% of arsenic contained in rice (Naidu, 2006).

Chapter 27

Status of the Codex Alimentarius Commission and Japan's Approach

Masahiro Segawa

Introduction

With the introduction of a risk analysis into the administration of food safety, Japan must promote a science-based administrative style. The Agreement on the Application of Sanitary and Phytosanitary Measures of the World Trade Organization (WTO) requires the member countries to base their domestic risk management measures on scientific principles and international standards. In response to these circumstances, the Ministry of Agriculture, Forestry and Fisheries (the "MAFF") formulated the standard work procedures for risk management relating to food safety[1] to steer the approach taken by the ministry.

Prompted by damage to health (by cadmium) and the inhibited crop growth (by arsenic or copper) in some areas that are highly contaminated by mine effluent, the Japanese government implemented remedial measures to combat heavy metal contamination such as cadmium. In addition, the Japanese government is well aware of international trends, including discussions at the Codex Alimentarius Commission (the "Codex"), established jointly by FAO and WHO for the development of international standards for food, and making every effort from the perspective of maintaining food safety, including conducing surveys of heavy metal content in crops, establishing and diffusing technologies for the suppression of cadmium uptake by crops, and, risk communication.

Status of the Codex alimentarius

1) Risk management of contaminants

The Codex Committee has a committee for each field, in which the Codex Committee on Contaminants in Foods (CCCF) has jurisdiction over toxic substances, such as contaminants like heavy metals, and mold that might be unintentionally introduced into food during production processesNote. The contaminant-related Risk assessments regarding the contaminants are under the jurisdiction of the Joint Expert Committee on Food Additives (JECFA) of

FAO and WHO, consisting of specialists in toxicological properties. The JECFA conducts toxicological assessments of contaminants and an evaluation of the intake from food based on requests from the Codex.

While the Codex is well known for their authority to establish international standards for contaminants in foods, they also focus on the prevention and reduction of contaminations throughout the course of crop production and the processing of foods so as to lessen risks the contaminants pose. Accordingly, the Codex developed the codes of practice for producers and processors of various contaminants to observe. While the establishment of the standards and the effects of eliminating non-compliant products from the market will primarily target the foods that contain a high concentration of contaminants, the implementation of proper technologies for reducing contaminants in the production or manufacturing processes will reduce the concentrations of contaminants in the regulated products in the market, and can ultimately achieve overall reduction in the distribution of intake of contaminants.

In addition to cadmium, which will be discussed later in this article, the Codex has investigated other environmental contaminants including lead, arsenic, dioxins and methyl mercury. The Code of Practice for the Prevention and Reduction of Lead Contamination in Foods has been established for lead, and standards have been set for concentrations of lead in agricultural, livestock and marine products. The Codex established the code of practice in respect of dioxins, but based on the recognition that the reduction of the concentration level in the environment at source would be more effective for the time and cost required by analysis, the Codex suspended the discussion on the standards. Although the JECFA sets a provisional allowable limit for the intake of highly toxic inorganic arsenic, their development work has been suspended since 1999 because the scientific morphology and morphological toxicity of arsenic in foods have yet to be elucidated, and there was no established analytical method for each form that arsenic takes in the body.

(Footnote)
Note: The Codex Committee on Food Additives and Contaminants (CCFAC) was in charge of this area until 2006. Since 2007, the CCFAC was split and reorganized into the Codex Committee on Contaminants in Foods (CCCF) and the Codex Committee on Food Additives (CCFA).

2) Discussion on a standard for cadmium

The Codex has been engaged in discussions of international standards for cadmium in foods based on the preliminary standards proposed at the 30th CCFAC meeting in 1998. In July 2005, the Codex adopted new standards for wheat and vegetables. The standards for polished rice and mollusks (e.g. saltwater bivalves and cephalopods) followed in July 2006. The 1998 draft preliminary standards included cereals, vegetables, fruits, meats (including internal organs), mollusks and crustaceans. Further discussions ensued based on the risk assessed by the JECFA, and the Codex discontinued discussions on food groups that were considered to have a lesser need for standards, taking into consideration regional contributions to cadmium intake. The standards that remained have also been modified according to actual cadmium concentrations.

Now, how will the Codex set the standards for contaminants, such as cadmium, which might be unknowingly contained in foods? The Codex General Standard of Contaminants and Toxins in Food (GSCTF) stipulates the following key principles:

(a) Only for those contaminants that present both a significant risk to health and a known or expected problem in international trade;
(b) Only for those foods that are significant for the total exposure of the consumer to the contaminant;
(c) According to the ALARA (As Low As Reasonably Achievable) principle.

The ALARA principle requires regulatory agencies to set the standard to the lowest possible limit to the extent reasonably achievable. Specifically, it sets the limits for contaminants in foods to a level that is slightly higher than the normal range of concentration, so as to avoid the unnecessary interruption of production or trade, on the premise that the health of consumers is protected, and that uncontaminated foods are produced by the application of appropriate technologies and measures.

The international standard for rice, the staple food for the Japanese, was 0.2 mg/kg in its original draft proposal of 1998. As the result of Japan's proposal based on various surveys of actual cadmium concentrations in rice and stochastic evaluation of intake[2] based on the surveys, the standard was revised to 0.4 mg/kg during the discussion process. The revision of the proposal was followed by the estimation by the JECFA of what impact the enforcement of these standards for various foods, including rice, might have on cadmium intake for each of the consumption forms over the world. Finally the Codex reviewed the results and adopted them as the standards based on the consideration of the results.

Japan's approach

1) Measures against cadmium

Japan experienced serious social problems arising from the adverse effects on the health of the residents who live in the areas severely contaminated by cadmium from mine effluent as a result of consuming water and rice grown in their own paddy fields. These problems prompted the government to restrict the distribution of rice under the Food Sanitation Act, and implement soil remediation projects under the Agricultural Land Soil Pollution Prevention Act.

At the same time, the government realized that the effect of cadmium on human health on which the Codex based its standards was intended to describe the health risk of ingesting food with the lower concentration than the above; a provisional allowable intake guideline was set by JECFA with a cadmium intake that would pose no health risk even if a person continued to ingest this amount for life. The limit was 7 μ g per 1 kg of body weight per week.

According to annual surveys on the intake of contaminants, which the MHLW has conducted since 1977, in 2004, a typical Japanese ingested an average 20 μg per day of cadmium from daily food, which was about 40% of the allowable intake set by the JECFA for a 50-kg person. Also, rice was the largest source of cadmium intake, accounting for one-half of the entire intake[3].

Based on these results, the MAFF began surveys of (concentration distribution) actual cadmium concentrations in mainly rice (approximately 37,000 rice samples) and other domestic agricultural and livestock products produced. The Ministry is currently developing and diffusing technologies for the reduction of risk at source during the production phase of the agricultural products. A technique of flooding the paddy fields around the time of heading of the rice, which provided the largest contribution to cadmium intake, to maintain moisture in the soil and absorb cadmium, has been found to reduce cadmium concentrations in rice. Farmers in the areas that might have the potential to produce rice with a high concentration are being actively encouraged to adopt this technique.

The water control method for paddy fields was implemented in 2004 in areas covering a total of 30,000 hectares, and more than 40,000 hectares in 2006[4]. A comparison with the fact that the areas highly contaminated by cadmium for which the soil dressing projects have been implemented over three decades since the 1970s covered 6,000 hectares[5] illustrates the scale of the project. Unlike engineering measures such as soil dressing, the effectiveness of the water control, a farming practice, is dictated by the weather conditions in that year. Yet, it is believed to be sufficiently effective in reducing the long-term intake of cadmium by consumers. A calculation of changes in the cadmium intake from food using data of the areas, where more than 0.4 mg/kg of cadmium was detected in the past from ongoing monitoring surveys by the MAFF, indicates declining trends since 2004. The MAFF is also involved in the development of technologies for the suppression of cadmium absorption by crops other than rice, phytoremediation, and soil cleansing.

2) Surveys on arsenic intake

The MAFF is in the process of compiling a priority list of toxic chemical substances to be controlled for risk management in order for the Ministry to be able to conduct systematic surveys on the status of these substances in foods[7], based on the information on food safety and the opinions of consumers and food industry sources.

The priority list includes environmental contaminants such as arsenic, cadmium, methyl mercury, dioxins and lead. For arsenic, lead and mercury, national surveys of domestically-grown agricultural products began in 2003 under a 4-year plan to collect basic data for discussions as to the need for future risk management measures. The overall results of these surveys are currently being compiled. According to the 2-year interim report, provisional calculations of the average intake by consumers from agricultural products indicate several conclusions[8]:

(a) Lead is less than 10% of the provisional allowable weekly intake set by the JECFA;
(b) Total arsenic intake is about 30% of the provisional allowable intake of inorganic arsenic assessed by the JECFA; and
(c) Total mercury is less than 10% of the allowable weekly intake of methyl mercury assessed by the Food Safety Committee for pregnant women.

As the concentration of arsenic in rice is higher than in other agricultural products, and the contribution of rice to intake is large, an analysis is being conducted on inorganic arsenic for which the provisional allowable intake has been set. In addition, studies and factual surveys regarding the analytical methods for arsenic in marine products in various forms are underway as these products are likely to be a large contributor to arsenic intake in Japan.

Food products are not significant contributors of dioxins. The baseline concentrations have been determined by national surveys during the period from 1999 to 2002. The baselines are used as a guideline for determining the effects of dioxins on agricultural products when environmental pollution occurs near agricultural lands. The MAFF is conducting a survey at the moment so as to determine whether yearly changes in the dioxin concentrations can be observed as a result of measures to control discharge of dioxins.

Conclusion

As stated at the beginning, Japan has just adopted the approach of risk analysis in the administration of food safety. For cadmium, the Pharmaceutical and Food Sanitation Council deliberates revisions to the domestic standards as soon as the results of current health risk assessment by the Food Safety Committee are announced. The Ministry must communicate the risks to consumers, producers and other stakeholders during the development of national measures for risk management. Although the MAFF is organizing discussion meetings in collaboration with the MHLW and disseminating information through the web sites of the ministries, it is essential that the information is scientific and appropriate (the probability and extent of adverse impacts on health) in regards to approaches that producers will implement with respect to risks if the communication is to be successful with all who are involved.

It is also important to implement the so-called food chain approach in the effort to secure the safety of foods. This approach not only controls the final products but also secures safety of foods through all stages from primary production to consumption. The MAFF expects to achieve this objective by developing a code of practice and a new framework, which will incorporate measures for the production processes in the Good Agricultural Practice technique (GAP) so as to promote them to the producers. In future, the MAFF plans to establish measures to control cadmium, such as the water control scheme, to be implemented at production fields through these frameworks.

As for arsenic, we are in the early stage of development of a risk management program. In conjunction with the current effort to determine the actual concentration levels in agricultural products, we must collect a wide range of scientific information, including data on toxicity to humans, and morphological changes and behavior of arsenic in soil and other production environments as well as in agricultural products.

1) Ministry of Agriculture, Forestry and Fisheries and Ministry of Health, Labor and Welfare (August, 2005), On Development of the Standard Procedures for Risk Management of Food Safety by Ministry of Agriculture, Forestry and Fisheries and Ministry of Health, Labor and Welfare.
2) Hiroshi Nitta (November, 2003), Study of Estimation of the Exposure to Cadmium by Japanese — The 2003 Interim Analysis Report
3) Ministry of Health, Labor and Welfare (August, 2006), Q & A on Cadmium in Foods
4) Ministry of Agriculture, Forestry and Fisheries (May, 2007), The 2007 Action Plan for

Measures Against Cadmium in Foods
5) Ministry of the Environment (December, 2006), Status of the Application of the Agricultural Land Pollution Prevention Act in 2005
6) Ministry of Agriculture, Forestry and Fisheries (July, 2007), Evaluation Results of the Agriculture, Forest and Fisheries Policy (Evaluation results of the measures implemented in 2006).
7) Ministry of Agriculture, Forestry and Fisheries (April, 2006), Mid-term Plan for Surveillance/Monitoring of Toxic Chemicals for Food Safety.
8) Ministry of Agriculture, Forestry and Fisheries (March, 2006), Interim Reports on Actual Contents of Lead, Arsenic and Mercury in Domestic Agricultural Products.

Chapter 28

Assessment of the Biological Effects of Cadmium Intake : Attempt at Estimation of Tolerable Intake

Hisayoshi Ohta

Biological effects of cadmium

One of the best known pollution diseases as the biological effects of cadmium (Cd) is the "itai-itai disease" which presents renal dysfunction and impairment of bone metabolism as a result of chronic exposure to cadmium. Reported biological effects resulting from acute and chronic cadmium poisoning at industrial work sites include pulmonary edema, metal fume fever, bronchitis, pulmonary emphysema and renal dysfunction. There have been reports, including the results of animal tests, that link cadmium to reproductive, urinary and cardiovascular diseases and diabetes, as well as potential carcinogenicity and endocrine disturbances (ICPS, 1992; Nogawa et al., 1999; Goyer, 1997).

The effects of chronic exposure to cadmium on renal function and bone metabolism are generally believed to be the onset of renal dysfunction followed by impaired bone metabolism. Problems appear to lie in the interpretation of indicators used for the assessment of renal dysfunction, and the determination of impaired bone metabolism in terms of osteoporosis and osteomalacia, as well as the relationship with the onset of these illnesses. The mechanisms for the transmission of cadmium through human cell membranes (i.e. a transport mechanism), the onset of renal dysfunction, and impairment of bone metabolism have yet to be elucidated in detail (ICPA, 1992; Nogawa et al., 1999; Goyer, 1997; Ohta, 2001; Ohta et al., 2000).

Issues relating to cadmium intake

There are some areas in Japan where a relatively high concentration of cadmium in agricultural products presents health concerns via exposure to cadmium through foods. These areas are the subject of ongoing epidemiological studies.

As rice is the staple food for the Japanese, an allowable intake level for cadmium in rice must be examined and established based on the dose-effect relationship of cadmium in the human body.

At present, the Food Sanitation Act prescribes 1.0 ppm to be the standard limit for cadmium concentration in rice in Japan. Distribution of unpolished rice that contains more than 0.4 ppm is prohibited. A standard of 0.2 ppm was proposed to the Codex Alimentarius Commission (the Codex), a joint international standards setting organization of FAO and WHO. The current standard in Japan is 0.4 ppm. Is the 0.2-ppm standard proposed by the Codex reasonable? Are there sufficient grounds for that standard? When it comes to the indicators for health assessment and significance levels and criteria for standards, a consensus has yet to be reached. The JECFA (the Joint Expert Committee on Food Additives of FAO/WHO) has set the provisional tolerable weekly intake (PTWI) at $7\mu g/kg/week$. Further experimental work for the assessment of biological effects of daily cadmium intake and the accumulation of epidemiological survey data are essential for risk assessment.

Issues relating to assessments of biological effects of cadmium

With respect to experimental work for an assessment of the biological effects of cadmium, there is a considerable pool of data accumulated from various experiments that used means such as injections of relatively large doses for the purpose of problem-solving or surgical means to elucidate the involvement of biological functions. There have been experimental studies using a prolonged administration of cadmium to test subjects through water or feed. However, few have investigated intestinal absorption in or assessment of biological effects of long-term exposure to cadmium by humans through daily intake. In particular, there has been no detailed examination of the effects of exposure to cadmium that take into consideration the normal physiological loads (pregnancy, childbirth and nursing) on female animals. There have been few detailed investigations into the modification of chemical forms during intestinal absorption and involving in vivo distribution and biological effects of cadmium.

Based on the results of the assessment of the biological effects of acute and subacute cadmium poisoning caused by usual environmental contamination or workplace exposure, the assessment of biological effects of cadmium at a daily intake level may be limited. In other words, the lower the concentration of cadmium is at exposure, the more affected is the modification of the biological effects of cadmium by the presence of metallothionein (MT), a metal-bound protein in intestinal tissues, modification of chemical form of cadmium derived by MT, and accompanying changes in biodistribution and the interaction of co-existing nutrient factors.

We need to conduct more detailed investigations into cadmium in terms of determining its intestinal absorption rate, developing proper assessment indicators for intake and accumulation, examining the load-modifying effects of pregnancy on female animals, mother-child transmission and effects of reproductive toxicity.

These investigations need to assess the biological effects of cadmium at the level of daily intake, and this is an important and current issue with respect to setting a tolerable intake (ICPS, 1992; Ohta, 2001; Ohta, H. et al., 2000; Ohta & Cherian, 1991; Rogers et al., 1997; Kovacs & Kronenberg, 1997; Brzoska et al, 1998; Bhattacharyya et al., 2000; Ohta et al.,

2006)

Attempt at an extrapolation of animal studies to humans and an estimation of tolerable intake

The results of an examination of renal function and bone metabolism of female rats orally administered with fixed doses of cadmium (2 – 60 mgCd/kg/day) showed that the onset of impairment of renal function and bone metabolism varied according to the conditions of cadmium exposure, suggesting that the impairment of bone metabolism was not an event secondary to the renal dysfunction but rather a direct effect of cadmium on bone metabolism. Based on this result, the relationship between the concentration of cadmium in kidneys relative to the dose in an animal study (Ohta et al., 2000) and the concentration of cadmium in renal cortex obtained by a study of clinical epidemiology on humans was examined, and an oral dose of cadmium with consideration given to the extrapolation to humans (Fig. 1) to examine the effects of reproductive loads and cadmium exposure on mothers. The bone metabolism of the mother showed a significant decrease in the density of femur owing to the effect of cadmium exposure in addition to nursing load. On the other hand, no significant differences were observed between the experimental group administered with cadmium in a dose equal to the estimated daily intake of humans and a control group (Fig. 2). The experimental group administered with cadmium at a dose twice the daily intake (7 – 8 μg/kg /day) showed significant loss in bone density, suggesting that the setting of a tolerable intake level of cadmium would need to take nursing load into consideration. The concentration of cadmium in the kidneys did not show any significant effects of pregnancy/nursing loads in this experiment with excretion of amino acids, NAG and β2MG into urine increasing significantly at 20 – 40 μg/g. This was an extremely low level of concentration compared to the conventional critical renal cadmium concentration of 200 μg/g.

With respect to this experiment, the author will present to the symposium the outcome of an attempt to estimate the tolerable daily intake of cadmium for humans based on the changes in various indicators for the assessment of abnormal renal function and bone metabolism obtained from the experiment, with aid of computer software provided by the US Environmental Protection Agency for the calculation of a benchmark dose (BMD).

Conclusion

As the biological effects of exposure to cadmium at a relatively high concentration in an industrial workplace or in a contaminated environment have sharply declined, the new question is how we should assess the biological effects of cadmium intake in our daily life. Is there an accumulation of study results sufficient to contribute to the setting of a tolerable intake standard for cadmium? Is the Japanese standard for cadmium reasonable in comparison with the results reported by studies overseas? Is it appropriate to assess the risk of an adverse effect of daily cadmium intake on health from the research results at a level of conventional environmental contamination? The issue is how and at what level should the indicators for adverse health effects of cadmium be set. There should be studies that are cognizant of more realistic junctions between experimental work using cells and animals and epidemiological studies using humans.

The Battle of the Codex: Standards and Food Safety ; The Case of Cadmium, Chloropropanol and Formaldehyde

Fujio Kayama

The standards for food distributed in world trade are set by the Codex Alimentarius Commission (the Codex). The Codex reports to the Joint Expert Committee on Food Additives (JECFA), a joint organization established by the Food and Agriculture Organization (FAO) and the World Health Organization (WHO). The JECFA is a group of experts who determine the total allowable intake for food additives and the tolerable intake for contaminants through foods and beverages. Based on the decision of the JECFA, the Codex Committee for Contaminants and Food Additives (CCFAC) sets the allowable concentration and tolerable concentration of a substance in each food item. The CCFAC is a subordinate body of the Codex. I attended six JECFA meetings: the 55th meeting in 2000, the 57th in 2001, the 61st in 2003, the 63rd in 2004, the 64th in 2005 and the 66th in 2006, as well as the related meetings: Codex Alimentalius meeting in 2002 and Committee for Contaminants and Food Additives (CCFAC) in 2003, as a technical adviser to the Government of Japan. Drawing from my experience, I would like to discuss the strategies of the US and EU for food safety and world food trade that emerged from these meetings.

The subject of this discussion is the contaminants that Japanese people ingest at high doses because of their love of food that contains these contaminants, as well as the contaminants in foods that are favorites of Asians but not popular among westerners. They are cadmium, methyl mercury, and chloropropanol.

At the 57th JECFA meeting, the spotlight was on the carcinogenicity of two of the chloropronanols, namely, 3-chloro-1, 2-propnanediol and 1, 3-dichloro-2-propanol (the "chloropropanols") which were generated as impurities during the acid decomposition of plant protein. Most of the intake from food sources comes from soy sauce. However, the concentration in soy sauce made by the traditional fermentation method in Japan should not create any problem, as this traditional method generates very little of those two substances. On the other hand, there may be some impact on Japanese producers of flavor seasonings.

Producers of oyster sauces in China, Southeast Asian countries and Korea may be required to change their manufacturing method, which may create a serious economic hardship.

Assessment reports of 3-chloro-1, 2-propanediol and, 3-dichloro-2-propanol can be found in papers published in the 1980s, but the studies were never on a large scale. The document that turned out to be the most important for assessment was not a published paper but a report prepared by Nestle. Study reports of a laboratory with Good Laboratory Practice are acceptable as assessment documents even if they have never been published. In spite of the fact that the Nestle report assessed the lowest dose as NOEL, the assessment by the Drafting Group of the JECFA produced a provisional maximum tolerable daily intake (PMTDI) that was lower than the Nestle's dose. The group assessed the minimum dose as LOEL and then multiplied it by a high safety factor. By the way, the JECFA Drafting Group consisted of Switzerland, the Netherlands and the United States.

This development suggested that the Drafting Group interpreted the study results to suit their agenda. These standards are not scientifically sound. Behind all this was the decision of EU standards on the same substances that was made public one week earlier. The intention of the EU representatives was apparently to harmonize the JEFCA standards with the EU standards. There is another point: this harmonization would be beneficial to Nestle. The company owns established processing techniques that use enzymes. They have commercial products out in the market place. The low standard might further freeze Asian flavor seasonings out of their markets.

During the meeting, the participants from Japan and Thailand pointed out that the assessment was scientifically wrong. Dr. Shubik of UK also pointed out errors in the argument. Yet, the opinion of a WHO committee member, who had the authority to rule, steered the argument toward the adoption of a 0 to $2 \mu g/kg$ body weight as PMTDI, as originally proposed by the Drafting Group.

The 66th meeting held in Rome in June 2006 discussed the draft proposal recommending that the PMTDI should be halved because study reports indicated that these substances might trigger adverse effects on the spermatogenesis of male rats born after in utero exposure. In the second week of the meeting, however, the reliability of the data analysis in that report was questioned, and the report was rejected. In the end, the current PMTDI was upheld. Although this deviates from the discussion about determining tolerable intake, our argument at the meeting was that "shoyu" made by the traditional fermentation method that contains little or no chloropropanols was the soy sauce to the Japanese people, and therefore, the term "soy sauce" should not be used in a Codex document. The discussions at the Codex meetings in the past indicated that we have had no chance of winning the argument. In the end, we concluded by having a category of "soy sauce containing protein that breaks down the subject acids" inserted in the beginning of the document.

The assessments of cadmium, dioxins and methyl mercury must also go through the process and steps unique to the JECFA, which are somewhat incomprehensible to first-time committee participants. Yet, some items such as the determination of safety and uncertainty factors are very difficult to understand by reading the assessment documents unless you attend the meeting in person to hear the discussion. I will explain the processes of the Codex assessment so you will understand how international standards are developed and adopted. As the standards are moving toward stricter levels, those who are involved in the standard setting should have a good grasp of the flow of the decision-making process.

In the background, there is an increasing demand on producer countries to set allowable standards at the lowest possible level according to the ALARA principle (As low as reasonably achievable), the Good Agricultural Practice (GAP) and the Good Manufacturing Practice (GMP) for the sake of keeping food safe. There are, however, large differences among the world's food cultures. There will be no objection to setting very strict standards for rice or soy sauce if such strictness has no impact on the food culture to which the committee members belong. The standards set by the JECFA will become the basis for the Codex standards. To the affected countries, the standards may create serious problems in their agriculture and trade.

The JECFA meetings are a battleground for scientific knowledge and logic for risk assessment experts. Attendees are required to participate and engage in discussions as individual researchers, not representatives selected by their government. You cannot deny the strong influences of the values based on your country and culture, particularly, the food culture. The JECFA meetings are a series of battles of scientific arguments complicated by an assortment of factors. Submission of documents, information and measurement data relating to assessments by the JECFA and the Codex in order for Japan to regularly send an army of experts to their meetings will be an international contribution that plays a pivotal role. Our participation in the past, however, was in the capacity of individual scientist without the support to wage a personal battle in the JECFA arena. Lack of preparation was a serious detriment. The United States often takes charge of developing the first draft. They start the preparation as early as three months before the meeting, and use several post-doc researchers working on each substance. The biggest contribution from Japan was standards for cadmium. A vast collection of screening data on cadmium concentration in rice by the MAFF was powerful ammunition. We must be well-prepared and vigilant in future so that we would not be put at a disadvantage in issues relating to food. The Japanese Government should be always prepared for the JECFA and the Codex.

Chapter 30

Heavy Metal Problems from the Perspective of Clinical Ecology

Kou Sakabe

Introduction

Impairment of health caused by heavy metals is an important issue in the filed of clinical ecology. In the United States, the American Academy of Environmental Medicine gives a high priority to the study of heavy metal poisoning, such as toxic metal syndrome (H.R. Casdorph and M. Walker) and chemical brain injury (K. H. Kilburn). The risk of health impairment due to exposure to heavy metals stretches wide covering fetuses to adults. The questions include fetal exposure through the mother's body, a possible linkage between exposure during the natal developmental stage of the central nervous system (the brain, in particular) and autism or attention deficit-hyperactive disorder (ADHD), and the generation of malignant tumors by chronic poisoning.

My presentation is entitled the "Heavy Metal Problems From the Perspective of Clinical Ecology"; I will overview the recent trends in heavy metal problems, and shed light on the problems of mainly arsenic exposure from the perspective of clinical ecology.

Background

Owing to the improvements in the environment for living and working in Japan and other advanced countries, we have less chance of being exposed to arsenic, and health problems are on a lesser scale compared to the past. From a global perspective, however, the risk of exposure to minute amount of arsenic still exists. Chronic arsenic poisoning induces disorders of the skin and peripheral nerve system, mental impairment, and hematopoietic disorder. In addition, arsenic is known to cause malignant tumors of the skin and internal organs (e.g. lung cancer, hepatic angiosarcoma, bladder cancer). Chronic arsenic poisoning caused by the contamination of well water by inorganic arsenic is well known, with incidences reported from many Asian regions, including China, India, Bangladesh and Thailand, as well as in South and Central America, including Mexico, Chile and Argentina.

Man's long relationship with arsenic

Our relationship with arsenic dates back to ancient Greece where arsenic was used as a medicine. Arsenic was also prized as a magic potion for eternal youth. Even today, a Chinese herbal medicine "xiong huang" is widely available at herbal medicine shops across China as an anti-inflammatory and antitoxic drug. Arsenic was also used in modern medicine from the 19th century to early 20th century, such as in Fowler's Solution. It was a panacea that cured scabies, syphilis, rheumatism and even cancer. One of the best sellers was arsphenamine developed by Hata and Ehrlich (commercial name: Salvarsan). The Ministry of Health, Labor and Welfare approved pasta arsenite (commercial name: Neoarsen Black) as a pulp devitalizer in dentistry, and a 0.1% arsenite solution for the treatment of leukemia in 2004. The latter is used for recurrent or intractable acute myelogenic leukemia. Outside medicine, arsenic is used in CCA (chromium, copper and arsenic) for the preservation of building foundations as well as in ant repellants as arsenite, Arsenic was also used as fertilizer until 1998. Arsenic sulfide is used as an additive to firecrackers.

Arsenic exists close to humans in a variety of things in our life, used as raw material for compound semiconductors, or additive to sheet glass to increase transparency. As food and drink, seaweed hijiki (hondawara family) contains a high concentration of inorganic arsenic. Wakame, kelp and nori also contain arsenic, but not much inorganic arsenic. Organic arsenic is more prevalent in fish and shellfish. The reader should be warned that some drinkable hot spring waters contain a high concentration of inorganic arsenic.

Arsenic poisoning

In what circumstances does arsenic poisoning occur in everyday life or at work? The most infamous food poisoning case was the Morinaga Arsenic Milk Case (1955). About 12,000 infants suffered subacute poisoning by arsenic that was accidentally mixed into milk stabilizer used in baby formula. More than 130 reportedly died as a result. Many victims still suffer from central nervous system disorders as an aftereffect. Air pollution was the culprit for arsenic poisoning in areas surrounding Toroku Mine and Sasagaya Mine, as well as cases in Quizhou province of China where coal fuel was the source. Arsenic causes serious occupational hazards in the workplace in the form of smelter fumes, dust from glass and the semiconductor industry. Areas of noteworthy environmental (natural) arsenic poisoning cases include China, Bangladesh, India, Thailand, Mexico, Chile and Agentina. The poisoning occurred from drinking contaminated water. A man-made environmental exposure to arsenic caused by an organic arsenic (containing diphenyl arsine) occurred in 2003 in the town (now city) of Kamisu in Ibaraki prefecture.

Mechanism of onset of arsenic poisoning

As described above, the chronic effects of arsenic have been known from human cases of chronic exposure to arsenic as well as through epidemiological surveys. There have been, however, many unknown factors regarding the mechanism of the onset of symptoms.

Mechanisms suggested as possible in the human body include the methylation and oxygen reduction of exposed inorganic arsenic that changes chemical forms rapidly. Establishing experimental animal species suitable for arsenic study is difficult, as metabolic rates vary widely from species to specie; also, in vitro experiments do not sufficiently reflect the biological effects.

It has been believed that humans reduced the toxicity of arsenic by methylating acutely toxic inorganic arsenic and eliminating it in urine, thus avoiding the impairment of health. Now we know methylated arsenic has a stronger involvement in carcinogenesis. In addition to carcinogenesis, some data suggest that methylated arsenic might be one of the culprits for benign skin symptoms by chronic exposure to inorganic arsenic. Furthermore, it is very important to know that loci of expression of symptoms differ according to the path of arsenic exposure. Occupational exposure leading to most cases of respiratory absorption is strongly linked to lung cancer, but if arsenic was absorbed through the digestive system, lung cancer is rare.

Biological reactions to arsenic are extremely complex. It will take considerable time to unravel the mystery completely. Yet arsenic poisoning by chronic exposure and trace arsenic contamination in ordinary environments are occurring globally right now. We must recognize that this is an urgent health issue.

Part V
Global Warming : Assessing the Impacts on Agriculture, the Environment, and Human Health, and Techniques for Responding and Adapting

31 **A Message from the Symposium Organizer** ························· 139
   Tadayoshi Shiba

32 **An Historical Overview of the GAIA Hypothesis
   and the IPCC Reports, and Global Warming in Japan** ················ 141
   Katsu Minami

33 **Assessment of Global Warming Impacts on Terrestrial Ecosystems,
   and Adaptive Techniques** ·· 146
   Yousay Hayashi

34 **Greenhouse Gases in Agricultural Ecosystems:
   Assessing Emission Rates and Developing Mitigation Technologies** ·········· 151
   Kazuyuki Yagi

35 **Health Impacts, Mainly Infectious Diseases, Due to Climate Change** ········· 156
   Hitoshi Oshitani

36 **The IPCC Now :
   Walking the Fine Line between Neutrality and Policy Prescriptiveness** ········ 159
   Anne Macdonald

37 **Climate Change Impacts, Adaptation, and Mitigation Measures:
Findings of the Synthesis Report** ·································· **160**
Hideo Harasawa

A Message from the Symposium Organizer

Tadayoshi Shiba

I would like to offer a few opening remarks on behalf of the sponsor of this Fifth Kitasato University Agromedicine Symposium.

It has been over two years since Kitasato University first communicated the concept of agromedicine. During that time we created the Agromedicine Committee and had the committee members prepare a document titled "On the Kitasato University Agromedicine Initiative."

To disseminate information we publish Kitasato University Newsletter of the President Office "Newsletter: Agriculture, Environment, and Medicine" each month, and have already reached issue number 36.

In the way of education, we started "lectures and seminars on agromedicine" in the School of Medicine and School of Veterinary Medicine. And this April in the College of Liberal Arts and Sciences we started a course called "Liberal Arts Seminar B: Agromedicine."

In the area of research, we set up "the development of standardization methods for determining the state of heavy metal ingestion and decreasing it," and we are promoting agromedical research that transcends colleges and involves the entire university.

To benefit society, we hold these Kitasato University Agromedicine Symposia in an effort to broaden agromedicine and provide scientific information. Two years have passed since starting these symposia, during which time we have addressed contemporary issues of agriculture and medicine through the environment under the themes "Agriculture, Environment and Healthcare," "Alternative Medicine and Alternative Agriculture," "A Look at Avian Influenza from the Perspective of Agriculture, Environment, and Medicine," and "Effect of Cadmium and Arsenic on Agriculture, the Environment, and Health." We also set up a system under which the results are made into pamphlets available for purchase by anyone.

This fifth symposium addresses the most relevant issue of our day: "Global Warming: Assessing the Impacts on Agriculture, the Environment, and Human Health, and Techniques for Responding and Adapting."

I would like to explain how we arrived at the main idea for this symposium.

Why is it that we do not give thought to the astonishing crises that humanity and civilization now face? Global warming is triggering extremely harmful phenomena in ecosystems, and Earth's biosphere has already surpassed the limit at which warming can be controlled. What is the reason that people nevertheless cannot understand this? Why is it that the United States backed out of the Kyoto Protocol, that developed and developing countries engage in political tug-of-war, and no progress is made on effective international measures to deal with global warming? Do we still lack an understanding, deep down in our hearts, of the concept that Earth is inhabited by all kinds of organisms, including human beings, from microorganisms on the small end to whales on the large end, and of the concept that these organisms are part of the "living Earth" that envelops an even greater diversity? Why don't people notice that all these crisis-like phenomena arise from our activities to produce a bounty of food and lead convenient and cultural lives? And even if they do notice, why can't they make any improvements?

But happily, in 2007 the Nobel Prize went to former US Vice President Al Gore, who had been working on the global warming issue since the 1970s, and to the Intergovernmental Panel on Climate Change (IPCC). This brought the global warming issue into the consciousness of people around the world.

The negative impacts on human livelihoods by global warming, which has become a crisis, are extremely serious. Food-related problems such as drought, salinization, and soil erosion, as well as medical problems like heatstroke, stronger UV radiation, dengue fever, and malaria all cast a dark shadow over the future of humanity. In any era, global environmental changes are intimately linked to agriculture, which provides us with food, and to medicine, which protects human health and life.

This symposium uses an approach from this perspective, and we invited speakers who have in various ways been active with the IPCC in Japan and other countries. I want to express my sincere gratitude to these speakers for kindly agreeing to appear today.

Chapter 32

An Historical Overview of the GAIA Hypothesis and the IPCC Reports, and Global Warming in Japan

Katsu Minami

Introduction

It was 1969 when everything changed. Like seeing oneself reflected in the surface of a river, in that year we first saw ourselves in the photographs of the blue Earth from the Apollo spacecraft. From that time we came to the realization that we cannot detach ourselves from Earth as a whole. And it seems we gained the subconscious awareness that perhaps Earth is one big living organism.

At the same time, 1969 was a creative year when Lovelock came up with the idea that Earth is the biggest organism in the solar system (the Gaia hypothesis). His hypothesis, announced that year, stated that the biosphere has a self-regulating function with the capacity to maintain our planet's health by regulating the chemical and physical environment.

Because the science and technology of the Apollo mission had developed consciously and rationally, we concentrated on seeing Earth as a whole from a bird's-eye view. This led to the formation of the Intergovernmental Panel on Climate Change (IPCC) and the participation of many scientists in climate change research, and this science and technology developed to the point of receiving the Nobel Prize.

On the other hand, while the Gaia hypothesis is also conscious and rational, one discerns an unconscious and intuitive background to a small degree. This hypothesis has had a major influence on many engineers and scientists in every field for approaching current problems concerning Earth. As a result, physicists, chemists, medical scientists, agriculturists, meteorologists, and many other scholars cooperated in integrating their knowledge. Further, this hypothesis even advanced into fields concerning the brain and spirit, with such names as "Global Brain" (Peter Russell) and "Earthmind" (Paul Devereux). These ways of thinking have made their way far and wide, now even integrating the knowledge of science and religion.

Therefore, in the historical background of global warming one can find conscious and rational aspects coexisting with unconscious and intuitive aspects. So here I would like to pursue "an historical overview of Gaia" and "an historical overview of the IPCC reports," and as time permits, describe some actual examples from "Global Warming: Impacts on the Festoon Islands of Japan." Further, we shall compile things that we can do right now, and consider what will happen to Japan in the way of global warming and culture.

An historical overview of the Gaia

The concept of Gaia was disseminated widely throughout the world by the British scientist James Lovelock, who graduated from college as a chemist, earned D.Sc. degrees in biophysics, Ph.D. degree in medicine, served as a professor in a medical college, and, as a consultant for NASA's space exploration plan, participated in a project to search for life on Mars. He is also an expert at gas chromatography, and the electron capture detector (ECD) he invented has contributed greatly to environmental analyses.

Lovelock explained that *Silent Spring* author Rachel Carson raised questions not as a scientist, but as an advocate, whereas he attempts to demonstrate the concept of Gaia, the living Earth, across a broad spectrum of scientific fields from astronomy to zoology.

Lovelock has issued many books on Gaia, including Gaia: *A New Look at Life on Earth, The Ages of Gaia*, Gaia: The Practical Science of Planetary Medicine, The Reveng of Gaia, etc. Last year, at age 89, he published The Revenge of Gaia: *Why the Earth Is Fighting Back – and How We Can Still Save Humanity*. In the Japanese-language version this was translated literally as *The Revenge of Gaia*.

In 1979, Oxford University Press published his book *Gaia: A New Look at Life on Earth*. This was published in Japanese translation in 1984. It took five years for the translation to be published.

In 1988, W. W. Norton published Lovelock's book *Ages of Gaia*. This appeared in Japanese translation in 1989. We were able to read this in Japanese translation only one year later.

Lovelock's recent work *The Revenge of Gaia* was issued in the original English and Japanese translation in 2006, so we were able to get the translation in the same year. The shortening time lag between the original and the translations of these 3 books bespeaks the strong interest that people have in Gaia. Further, the warming of Earth also raises people's interest in the biosphere.

Twenty-seven years passed between the publication of *Gaia: A New Look at Life on Earth* and the appearance of *The Revenge of Gaia*. That is over a quarter century. To sum up *Gaia: A New Look at Life on Earth* in a few words, it is proof of the hypothesis that Earth's organisms, atmosphere, oceans, and soil make up a complex system that can be seen as a single organic whole, and has the capacity to maintain our Earth as a place suitable for life.

The Ages of Gaia was totally rewritten on the basis of new scientific knowledge that emerged after *Gaia: A New Look at Life on Earth* was written. Nine years elapsed in that time.

In the introduction, Lovelock emphasizes that he only wants to speak for Gaia because there are far fewer people speaking for Gaia than speaking for humans. In "The Hippocratic Oath," he explains one of this book's purposes by saying that a specialized field called planetary medicine is needed, and that as its foundation it is necessary to create geophysiology.

Especially notable about this book is this passage, which some time ago forecast an IPCC conclusion: "The health of the Earth is most threatened by major changes in natural ecosystems. Agriculture, forestry, and to a lesser extent fishing are seen as the most serious sources of this kind of damage with the inexorable increase of the green-house gases, carbon dioxide, methane, and several others coming next."

"There is no way for us to survive without agriculture, but **there seems to be a vast difference between good and bad farming.** Bad farming is probably the greatest threat to Gaia's health."

In *The Revenge of Gaia*, Lovelock explains that Gaia is attempting to eject human beings. He also says that there are too many human beings for Gaia to accept, and he writes that until nuclear fusion and hydrogen energy technologies become viable, the electricity that is the basis for supporting the oversize human population will have to come from nuclear fission, which has the smallest environmental burden.

Lovelock says the critical point for global warming is a CO_2 concentration of 500 ppm. If the amount of Arctic ice that melts increases, he says, the CO_2 trapped in the ice will be released and further spur on warming. Here we catch a glimpse of the threshold value issue, which people do not talk about much. One must be aware of the atmospheric CO_2 concentration, and the threshold value, which is determined by air temperature. Once this value is exceeded, the result cannot be changed no matter what one does. Earth has reached an unprecedentedly high temperature, and it is too late to go back.

Tuvalu, in the South Pacific, is faced with the crisis of inundation. Will it be necessary for Japan to build sea walls to prevent the inundation of coastal plains due to expansion of seawater caused by rising air temperature? When Earth rapidly heads toward a new and sweltering state, climate change will doubtless bring chaos to political and business circles.

An historical overview of the IPCC reports

The origin of the IPCC goes back to the start of research on the climate and climate change by WHO and UNEP, occasioned by worldwide extreme weather events such as heavy flooding, drought, and warm winters. As international challenges related to climate change increased, there was a rising need to comprehensively provide scientific information on climate change so that governments would devise effective policies. Against this backdrop, ideas for creating the IPCC were proposed at the 1987 WMO Congress and the UNEP Governing Council meeting. Approval and establishment of the IPCC came in 1988.

Although the IPCC was originally founded without any connection to the United Nations Framework Convention on Climate Change (UNFCCC), because its First Assessment Report was considered outstanding for compiling and assessing knowledge on climate change, it came to be used widely as a basic reference.

The First Assessment Report (1990) includes: Scientific Assessment of Climate Change — Report of Working Group I [WGI] (increases in greenhouse gases and their contributions to global warming are important), Impacts Assessment of Climate Change [WGII], The IPCC Response Strategies [WGIII], Climate Change: The IPCC 1990 and 1992 Assessments, and summaries for policymakers (SPMs).

The 1994 IPCC Special Report (Climate Change) includes Radiative Forcing of Climate Change and An Evaluation of the IPCC IS92 Emission Scenarios, IPCC Technical Guidelines for Assessing Climate Change Impacts and Adaptations, IPCC Guidelines for National Greenhouse Gas Inventories, and the 1994 Special Report SPM.

The Second Assessment Report, Climate Change 1995, includes The Science of Climate Change [WGI], Impacts, Adaptations and Mitigation of Climate Change: Scientific–Technical Analyses [WGII] (technologies for greenhouse gas reductions are important), Economic and Social Dimensions of Climate Change [WGIII], IPCC Second Assessment Synthesis of Scientific–Technical Information Relevant to Interpreting Article 2 of the UNFCCC, and Summaries for Policymakers of the three Working Group reports.

The Third Assessment Report, Climate Change 2001, includes The Scientific Basis [WGI], Impacts, Adaptation & Vulnerability [WGII], Mitigation [WGIII], and the Synthesis Report.

The Fourth Assessment Report, Climate Change 2007, includes Technologies, Policies, and Measures for Climate Change Mitigation; Description of the Simple Climate Model Used in the IPCC Second Assessment Report; Atmospheric Greenhouse Gases: Physical, Biological, and Socioeconomic Impacts; Impacts of Proposals for Restricting Carbon Dioxide Emissions; and Climate Change and Biodiversity.

IPCC Special Reports include: The Regional Impacts of Climate Change: An Assessment of Vulnerability (1997), Aviation and the Global Atmosphere (1999), Methodological and Technological Issues in Technology Transfer (2000), Emissions Scenarios (2000), Land Use, Land–Use Change, and Forestry (2000), Guidance Papers on the Cross Cutting Issues of the Third Assessment Report, and Climate Changes the Water Rules: Dialogue on Water and Climate Synthesis Report.

Global warming impacts on the festoon islands of Japan

Here are some impacts of global warming on the festoon islands of Japan: Melting of permafrost (Mt. Fuji and Hokkaido); poor catches of sardines (Sardinops melanostictus) (off coast of Sanriku region); damage to rice crops (Kyushu); sea level rise (western Japan); coral damage (Okinawa); deer winter in mountain (Tochigi, Gunma, and Hokkaido prefectures); decline of alpine plants (beech forests, Callianthemum miyabeanum); reduced apple harvests (Aomori Prefecture); larger methane emissions (rice fields); northward migration of the cicada Cryptotympana facialis (Tokyo); more tropical nights (Tokyo); large outbreaks of Nomura's jellyfish (Nemopilema nomurai) (coastal areas of Japan Sea and Sanriku region); enlargement of Asian skunk cabbage (Lysichiton camtschatcense) (Oze); wetland loss (Kushiro); reduced transparency of lake water (Lake Mashu); threat of extinction for the Okinawa rail (Gallirallus okinawae) (Okinawa); loss of sand dunes and coastal erosion

(Shizuoka and Chiba prefectures); major outbreaks of cyanobacteria (Ibaraki Prefecture); unusual changes in urban vegetation due to warm winters (Tokyo); and reduced numbers of the ptarmigan (*Lagopus muta*) in the Southern Japan Alps.

What we can do right now?

Why do our thoughts not extend to the extraordinary crises that humanity and civilization now face? Heating due to global warming is bringing about phenomena which are extremely harmful to ecosystems, and has already exceeded the level at which warming could be controlled, yet why can't people understand that?

What can we do right now? Return carbon and nitrogen to the soil. Devote ourselves body and soul to buying green products. Decrease our desire for material things. Reduce consumption of resources and energy. Reduce pollution and waste across the board in production, distribution, and consumption. Turn our attention not only to CO_2, but also to CH_4 and N_2O. Though we might criticize politicians, the media, and the national structure and system, we need the self-awareness that we ourselves cause global warming. Transition to and lead the way to a negative-growth economy. Cultivate thinking in which the economy is a subset of the environment, and immediately discard thinking in which the environment is a subset of the economy.

Global warming and culture

Won't global warming also affect the culture of beautiful Japan? Might we not expect, for example, the loss of beautiful landscapes, the loss of communities in residential areas, qualitative changes in customs, the loss of biodiversity, changes in music and poetry, and a crisis in the mental world, which is not a matter of numerical values shown scientifically? There is little doubt that we too will experience phenomena similar to the tragedy of Bhutan, where glacial lake outburst floods occur soon.

Chapter 33

Assessment of Global Warming Impacts on Terrestrial Ecosystems, and Adaptive Techniques

Yousay Hayashi

People first became aware of the global warming crisis at the 1985 Villach Conference, which adopted a declaration saying that the increase in world temperature in the first half of the 21st century will probably be a substantial one that humanity has never before experienced. Over 20 years have passed since global warming became an internationally critical issue, and now the global ecosystem is partaking of the "reality" predicted in that declaration. In recent years an international framework of the Intergovernmental Panel on Climate Change (IPCC), conferences of the parties (COP) to the United Nations Framework Convention on Climate Change, and other institutions has been established to tackle the problem of global warming, and in 2005 the Kyoto Protocol, which specifies numerical targets for greenhouse gas reductions, came into effect.. As we enter 2008, the year when the Kyoto Protocol's First Commitment Period starts, we shall see what real effect the measures to combat global warming have.

The need to reduce greenhouse gases is substantiated by the impacts of past and projected global warming. To summarize the Working Group I report of the 2007 IPCC Fourth Assessment Report, "The Physical Science Basis," the results of observations made in recent years show: (1) Warming is unequivocal, and there was an increase of 0.74 °C in the global average temperature over the 100 years of observations from 1906 to 2005, (2) the average world sea level increased at the annual rate of 1.8 mm from 1961 to 2003, and there is a very real possibility that the melting of the Greenland and Antarctic ice sheets is contributing to this, (3) mountain glaciers and snow cover have dwindled in both the northern and southern hemispheres, (4) the frequency of heavy rain has increased over most land areas, and (5) there are fewer cold days and frosts, while there is a greater frequency of hot days and heat waves. Based on these facts, the report says, "Most of the observed increase in global average temperatures since the mid-20th century is very likely due to the observed increase in anthropogenic greenhouse gas concentrations," thereby concluding that the increase in anthropogenic greenhouse gases is the cause of global warming.

On projections for the future, the report offers the results of warming projections based on a number of social scenarios called SRES scenarios. The main predicted changes are as follows: (1) Temperature: Over the next 20 years temperature will rise at a rate of about 0.2 ℃ per decade. Average world temperature at the end of the 21st century will be 1.1–6.4 ℃ higher than the annual average between 1980 and 1999. (2) Oceans: The normal value of the global average sea level from 2090 to 2099 will be 0.18?0.59 m higher than the normal value for 1980 to 1999. At the same time, the increased atmospheric carbon dioxide concentration will cause acidification of the oceans, and at the end of the 21st century the global average pH of ocean surface water will be 0.14–0.35 lower. (3) Precipitation: It is very likely that average annual precipitation will increase in high latitudes. On the other hand, in many subtropical regions average annual precipitation will decline (it is likely that the normal value for 2090–2099 will be a maximum 20% lower than the normal value for 1980?1999). (4) Changes in the polar regions: Major changes are predicted. In spite of differences among greenhouse gas emission scenarios, sea ice will shrink in both the Arctic and Antarctic. In particular, Arctic ice in the late summer will almost totally disappear by the second half of the 21st century.

Temperature and precipitation in East Asia are reported as follows: Average annual temperature is projected to rise in comparison with the normal value for 1961–1990. The normal value for 2010–2039 will be 1.3–1.5 ℃ higher, that for 2040–2069 will be 2.5–3.6 ℃ higher, and that for 2070–2099 will be 3.4–6.1℃ higher. Precipitation is likewise projected to increase in comparison with the 1961–1990 normal value. The average value for 2010–2039 in winter (December through February) will be 5%–6% higher, that for 2040–2069 in winter will be 10%–13% higher, and that for 2070–2099 in winter will be 15%–21% higher.

Terrestrial ecosystems have in the past been affected in various ways by rapid changes in environmental conditions, and it is predicted that they will also be affected in the future. The IPCC Fourth Assessment Report (Working Group II) can be summarized as follows:

• Flora and fauna are migrating toward the polar regions or to higher altitudes. Responses to global warming by various terrestrial species are being manifested as changes in growth stages (phenological changes), especially the earlier occurrence of springtime phenomena and bird migrations, and as lengthened growing seasons. Satellite images taken since the early 1980s show that in many regions vegetation greens earlier in spring, and that because the growing seasons of plants are longer, their net primary productivity has increased. Although there are still few cases, we are seeing the disappearance of endemic species and changes in species composition over the last 20 to 30 years.

• The impacts of global warming on agriculture and forestry are still limited compared to other fields, but earlier crop growing seasons are clearly observed across broad swaths of the northern hemisphere. In particular, changes in crop management such as earlier planting have appeared in the higher latitudes of the northern hemisphere. In many regions, longer growing seasons are manifested as increased forestry production. Meanwhile, higher temperatures and dry conditions in other areas have combined to cause declining forestry productivity and forest fires. Agriculture and forestry have weak defenses against recent heat waves, drought, and floods.

• By 2100, ecosystems will be exposed to the highest atmospheric CO_2 concentration in the last 650,000 years, and to the highest average global temperature level in the last 740,000 years. Seawater pH will be the lowest in the last 20 million years. The combination of unprecedented changes, such as disturbances related to climate change (floods, droughts, forest fires, insect and disease outbreaks, seawater acidification, etc.) and other global-scale changes (land-use changes, population growth, excessive resource extraction, etc.), is expected to exceed the capacity of many ecosystems to adapt. When environmental conditions exceed a certain threshold of ecosystem restoration capacity, they are likely to trigger changes which cannot be undone.

• Currently the terrestrial biosphere serves as a carbon sink, but it is thought that its effectiveness will peak and then start declining about mid-century, whereupon it will become a carbon source and aggravate climate change. An example of this is the accelerated emission of methane from the tundra. At the same time, the buffering capacity of the oceans will be saturated. Under such conditions, the concentration and emission of CO_2 will be higher than at present. Land-use changes and logging of rainforests are among the factors affecting the global carbon balance.

• Scientists think that the danger of extinction will increase if the average global temperature exceeds that since the Industrial Revolution by 2-3℃; so far 20%-30% of all species have been evaluated. Compared with conditions in the geological past, the danger of species disappearance is the highest ever.

• A temperature increase of 2-3℃ over that at the time of the Industrial Revolution and the level of atmospheric CO_2 concentration underlying that increase would probably induce intrinsic changes in the structures and roles of terrestrial and marine ecosystems.

• Results of model experiments performed on warming regions found that if the average temperatures of farming regions have moderate increases (1-3℃), then because of the concomitant increase in CO_2 concentration and changes in amount of rainfall, grain yields would increase somewhat, thereby having a positive effect. But in low-latitude zones, especially tropical areas with dry seasons, many grains would yield less even at a temperature rise of 1-2℃. That would increase the danger of famine. A further temperature increase would have a negative effect on all regions.

• Climate change will raise the number of people exposed to the danger of famine, which would in turn substantially hinder socioeconomic development.

• It is clearly predicted that, in addition to warming occurring as a long-term trend, the increased frequency and intensity of extreme weather events will result in unstable food and forestry productivity. Recent research suggests that frequent heat waves, droughts, and floods will eclipse the impacts of warming and have detrimental impacts on crop yields and livestock production. In other words, there are concerns that negative impacts will be greater and come sooner than impacts under the average situation alone.

• As temperatures rise, water will be strictly managed. However, for temperature increases in the range below the medium level, it is predicted that appropriate adaptation measures will result in many advantages. The effects of adaptation will be varied, ranging from only slightly alleviating negative impacts to changing negative impacts into positive ones. In grain cultivation systems, for example, adaptation such as changing cultivars or planting times could avoid 10%–15% of predicted yield declines. If converted to a temperature difference in a certain region, this corresponds to a temperature increase of 1–2 ℃ (in equivalent to a decrease of yield). Changing policies and systems is essential to facilitate adaptation. Adaptive techniques should be implemented in coordination with development tactics, local policies, strategies for eliminating poverty, and other such efforts.

• Recent re–analysis of FACE (Free Air CO_2 Enrichment) experiments found that at a 550 ppm concentration C3 plants realize a 10%–20% yield increase in the absence of other stresses. C4 plants had a 0%–10% increase. Estimates using a crop model under conditions of heightened CO_2 concentration provided results that matched these experimental values. Recent FACE experiments found no definite productivity response in grown forest communities, but found that growth was augmented in young trees.

The IPCC reports bring together a variety of assessments on the impacts of global warming on ecosystems, and it is important to consider this matter as an urgent task from the following perspectives: First is that when multiple influences come to bear simultaneously, there are unexpected effects; second, unless one looks at the widest diversity of ecosystems possible, one will not discern the true situation. An example of the first would be the fertilization effect of raised CO_2 concentration on grains (a favorable effect) actually is manifested as fertility decline (a negative effect) when it acts at the same time as temperature rise. In an example of the second, while greater insect damage due to global warming is predicted, the effect on a simultaneous increase in predators that feed on those insects could have some impact on the number of insect pests.

It is of the greatest importance to use the situations, future projections, and assessments discussed above in mitigating global warming impacts. In this connection, it is very important to ascertain and assess global warming impacts as risks. This is related to Article 2 of the Framework Convention on Climate Change, which states, "The ultimate objective of this Convention and any related legal instruments that the Conference of the Parties may adopt is to achieve, in accordance with the relevant provisions of the Convention, stabilization of greenhouse gas concentrations in the atmosphere at a level that would prevent dangerous anthropogenic interference with the climate system." This article points out the importance of analyzing the threshold at which there is a danger level of warming and impacts, and shows the need to clear up uncertainty and to perceive impacts as risks.

While it is essential that governments abide by the Kyoto Protocol's numerical targets, due to the great inertia on the limiting of atmospheric CO_2 emissions, it is thought that for the time being it will be impossible to arrest the global rise in temperature. Therefore, we need very much to provide answers for how to make ecosystems adapt, and what technologies there are for mitigating negative impacts at levels of stress caused by a certain degree of global warming.

Chapter 34

Greenhouse Gases in Agricultural Ecosystems: Assessing Emission Rates and Developing Mitigation Technologies

Kazuyuki Yagi

Introduction

Agriculture is a necessary occupation of humanity which makes the maximum use of an ecosystem's energy and material balances. Whether it is primitive swidden agriculture or mechanized intensive agriculture that applies chemical fertilizers and pesticides, agriculture modifies the various cycles in ecosystems and transitions the energy and material equilibrium that has been maintained for a long time to another balanced state. While this made possible the provision of materials such as food and fiber that are civilization's foundation, it has also caused problems that modern humanity now faces, such as the chemical burden on the environment and changes in the hydrological cycle and energy balance.

One such problem that has been found is the emissions of greenhouse gases from agricultural ecosystems, which is noted to be a cause of current global warming. According to the IPCC Fourth Assessment Report (AR4)[1], which was released last year, the amount of greenhouse gases generated by the world's agricultural ecosystems is 5.1–6.1 Gt CO_2 eq (carbon dioxide equivalent), accounting for 13.5% of anthropogenic emissions. The emissions and absorption of CO_2, the greatest of these greenhouse gas emissions, are thought to be in balance globally. But because the greenhouse gas emissions from forests, which are calculated separately, include emissions caused by changing land use to farmland, one could say that the impacts of agriculture extend also to the forestry sector. Additionally, agricultural ecosystems account for over half the anthropogenic emissions of two other greenhouse gases, methane (CH_4) and nitrous oxide (N_2O), and therefore are major sources.

Greenhouse gas emissions and mitigation technologies

CO_2 emissions from farmland

Whether farmland is a carbon sink or source is determined by the balance between the absorption of atmospheric carbon dioxide by crop photosynthesis on the one hand and emissions by crop respiration and the decomposition of soil organic matter and crop residues on the other. Tilling farmland accelerates the decomposition of organic matter in the soil, and carbon is often taken out of the system as the harvest. Thus, soil organic matter, which had accumulated in a state of equilibrium when the land was forest or grassland, declines when the land is tilled, and tends to be emitted as carbon dioxide. However, inputs of organic matter such as compost and livestock waste slow that decline, and the accumulation as soil organic matter increases. In other words, sometimes there is a carbon storage effect.

In addition to the application of organic matter such as compost and livestock waste, other agricultural technologies that are effective ways of storing carbon are no-till and reduced tillage, crop rotation, and cover cropping. Worldwide, it is also important to curb the conversion of forests and wetlands to farmland. According to IPCC AR4[1], there are considerable expectations for the ability of farmland soil to store carbon. It is estimated that there is a mitigation potential of 3870 Mt CO_2 eq by 2030 at carbon prices of up to 100 US\$ per t CO_2-eq, which corresponds to about 8% of the anthropogenic greenhouse gases emitted in 2004.

CH_4 from rice fields[2]

Flooding of rice fields by irrigation creates an anaerobic environment in which the activity of a group of obligate anaerobic archaebacteria called methanogens synthesizes methane as an end product of organic matter decomposition. This is released into the atmosphere. Global annual methane emissions from rice fields are estimated to be 20-100 Tg, accounting for 5% -30% of anthropogenic emissions. On-site measurements of rice field methane emissions have been carried out since the 1980s, and currently results from over 100 locations, primarily in Asia, have been released. The results have been compiled, and the 2006 IPCC Guidelines[3] call for a baseline emission factor (130 mg m^{-2} day^{-1}), and a scaling factors for increased emissions due to water mangement and organic soil amendments. In the early 1990s, a nationwide study performed on Japan's rice fields estimated annual methane emissions to be 330,000 t.

Suggested techniques to reduce rice field methane emissions include water management by using midsummer drainage and intermittent irrigation, organic matter management by which rice straw is composted and decomposition is encouraged when rice fields are not flooded, use of fertilizer or agricultural supplies, and soil amendments. Many of these are proved to be effective. Under a program of the Ministry of Agriculture, Forestry and Fisheries starting in 2008, the ministry is planning the nationwide demonstration and promotion of techniques to reduce methane emissions by means of water management for the purpose of helping Japan meet its greenhouse gas emission reductions for the First Commitment Period under the Kyoto Protocol.

N_2O emissions from fertilizer nitrogen[4]

The nitrogen needed for crop production that is applied to farmland as chemical fertilizers or organic material changes form in the soil due to the action of microorganisms, so that NH_4-N becomes NO_3-N (nitrification) and NO_3-N becomes N_2 (denitrification). N_2O is formed as a byproduct in the soil in the processes of nitrification and denitrification, and is released into the atmosphere. The amount of N_2O emitted from farmland generally increases with the amount of nitrogen applied, and an emission factor, which is the proportion of N_2O-N formed per amount of nitrogen applied, is used to estimate the amount of N_2O emitted. The 2006 IPCC Guidelines[3] suggest 1.0% as the standard emission factor (default value). However, data from observations in Japan in many cases have an emission factor that is lower, while in a few cases such as tea field soil, extremely high emissions have been observed. Based on these research results, annual N_2O-N emissions from Japan's agricultural land are estimated to be 4420 t N eq. The N_2O emission processes are observed to be direct emission into the atmosphere from farmland, as well as indirect emissions in which nitrogen applied in farming areas runs off into groundwater and rivers, after which N_2O is released by degassing.

As N_2O emissions from farmland soil have much in common with another environmental problem, that of nitrate nitrogen leaching into groundwater, it is effective to develop mitigation techniques which emphasize, for example, making crops use nitrogen more efficiently, and suppressing nitrification and denitrification. Some conceivable techniques for this are improvements in fertilizer application methods, such as application of optimum amounts of nitrogen, and split or localized application; using new types of fertilizer such as slow-release fertilizer, nitrification inhibitors, and urease inhibitors; and the appropriate application of organic matter. It requires effort to use these techniques and to offer and widely disseminate nitrogen application systems appropriate to each geographical region to maintain high yields without exceeding the soil's environmental capacity; however, it is necessary in order to harmonize food production and environmental conservation.

CH_4 and N_2O emissions from the livestock industry[5]

Methane from ruminant livestock, and methane and N_2O from livestock waste are major sources of agricultural greenhouse gases. Global annual methane emissions from ruminants are estimated to be 80-90 Tg, which corresponds to about 25% of anthropogenic emissions and is believed to be the largest source of methane emissions. Meanwhile, although there is great uncertainty in the estimated emissions from livestock wastes, they are without a doubt a major source. Like rice fields and fertilized soil, these two sources arise from the activity of microorganisms, and generate methane and N_2O by the same mechanisms.

In the first stomach (rumen) of cattle, goats, sheep, and other ruminants, methane is constantly generated from feed carbohydrates, and its amount usually corresponds to 2%-12% of ingested energy. Because emitting methane represents an energy loss for livestock, research into mitigating emissions is also being carried out from the pint of view of using feed more efficiently. But because methane synthesis also functions to eliminate metabolic hydrogen, which is harmful to the proliferation of microorganisms in the rumen, appropriate control is needed. With this in mind, one recognized effective way of both improving livestock productivity and diminishing methane emissions is administering preparations (such as copper sulfate, ionophores, antibiotics, and unsaturated fatty acids) that control the activities of protozoa and microorganisms in the rumen. It is also effective to reduce high-fiber feed (roughage) and increase the proportion of concentrates, which have high nutritive value. As a more practical applicable technique, developing countries have confirmed the effectiveness of adding food manufacturing byproducts such as rice bran and brewer's grain.

There is a variety of conceivable ways to hold down emissions from livestock waste; these involve feed, waste disposal (composting), applying waste to farmland, and the like. But for composting and application to farmland such measures are complicated because of the frequent tradeoff arising with methane generated under reductive conditions, and N_2O generated due to more oxic conditions. Recently it has been observed that adding essential amino acids could improve the efficiency of feed use and reduce the amount of livestock waste generated per head or per unit product, raising expectations for this as a way to mitigate greenhouse gas emissions.

Possibilities for reducing greenhouse gas emissions from agricultural ecosystems

As seen above, various sources in agricultural ecosystems emit the three major greenhouse gases CO_2, CH_4, and N_2O. Quantitative assessments of local and global emissions contain uncertainties owing to the diversity of agricultural ecosystems, and although there is room to reduce those uncertainties, it is clear that there is significant global warming impact. Many techniques have been proposed to mitigate emissions from farmland and livestock, and on-site tests and other procedures have confirmed the considerable reductions achieved by many of these techniques, such as water management (in rice fields), organic matter management, and fertilizer management for farmland, and feeding and waste disposal management for livestock. However, at this time there are very few instances in which these techniques have been implemented in actual farming operations to reduce greenhouse gas emissions. Under the Kyoto Protocol, only Canada and three other countries have chosen farmland as carbon sinks. And although some countries are planning CH_4 and N_2O reductions, it seems there are still no cases of actual implementation.

One reason that techniques for reducing greenhouse gas emissions from agricultural ecosystems have yet to be put into practice is the insufficient assessment of the economic benefit conferred by cutting emissions. Taking into account cost and labor, since many farming operations are family-run, there is hardly any possibility for wide adoption of a technique unless it improves overall earnings and makes work easier. This makes it necessary to conduct detailed assessments of each technique's economic effect in various regions, and to show the possibility of it being accepted by farmers. Also needed is policy support to promote such techniques.

Additionally, although emission-mitigating techniques have been assessed for farmland and livestock themselves, further efforts will be needed to counter the insufficiency of assessments covering whole agricultural ecosystems and entire geographical regions. Dealing with this problem will require the introduction of lifecycle assessment (LCA) of the new techniques that look at the balances of whole systems and include such factors as additional inputs of energy and agricultural supplies or the production and use of feed. Take for example livestock wastes, which have the potential to release greenhouse gases at various times during their handling. It is necessary to comprehensively assess area-wide collaboration between livestock farmers and crop farmers and find the solution which best satisfies both productivity and farming economy, while at the same time having the smallest greenhouse gas emissions and not increasing other environmental burdens. For biofuels, whose production is growing in recent years, researchers should also assess their effectiveness in reducing fossil fuel use, and the possibility that their cultivation will increase greenhouse gas emissions.

Another issue is that the proportion of greenhouse gas emissions from the agriculture sector in developed countries is relatively low, putting most of the reduction potential in developing countries. In Japan, agriculture accounts for a mere 2% of the greenhouse gas emissions inventory, while in tropical Asia, whose agricultural system is also based on wet rice cultivation as in Japan, agriculture accounts for a very high proportion of emissions, such as 28% in India and 35% in Thailand. Especially in countries with vast areas of farmland and many domestic animals, measures to reduce emissions by implementing agricultural techniques could make a large contribution. It is predicted that in such countries, attuning the implementation of these techniques with sustainable development policies would further advance the possibilities for reductions. It is possible that the clean development mechanism (CDM) provided for by the Kyoto Protocol could be used as a new development assistance tool.

IPCC AR4[1] shows clearly that, in terms of cost, greenhouse gas emission reduction measures in the agricultural sector can compete with those in the energy, transport, forestry, and other non-agricultural sectors. An advantage cited is that one can expect long-term effectiveness, and that overall a major contribution could be made. Developing new techniques in the agricultural sector is not only a promising way to reduce greenhouse gas emissions, but also coincides with the orientation toward sustainable or environmental friendly agriculture, which is the desirable future form of agriculture. Certainly now when we are pressed to act on global warming, we perhaps have a good opportunity to press forward with international negotiations to make appropriate land use possible, and to blueprint the desirable form of future agriculture that is in harmony with nature.

References

1) IPCC (2007): IPCC Fourth Assessment Report (AR4): Climate Change 2007, Cambridge University Press. http://www.ipcc.ch/
2) Yagi, Kazuyuki (2004). *Atmospheric Methane Dynamics and Methane Emissions from Rice Fields*. "Agro-Environmental Research Series, No. 15, Carbon and Nitrogen Cycles in Agricultural Ecosystems," pp. 23-50, National Institute for Agro-Environmental Sciences.
3) IPCC (2006): IPCC Guidelines for National Greenhouse Gas Inventories. http://www.ipcc.ch/ipccreports/methodology-reports.htm
4) Yagi, Kazuyuki (2006). "Greenhouse Gas Emissions and Their Assessment." *Fertilizer Encyclopedia*, pp. 358-365, Asakura Shoten.
5) Japan Livestock Technology Association (2002). *Controlling Livestock Greenhouse Gas Emissions* (Compendium). http://jlta.lin.go.jp/kokunai/houkoku_jigyo/h13_op01.html

Health Impacts, Mainly Infectious Diseases, due to Climate Change

Hitoshi Oshitani

In 2003, international organizations (WHO, WMO, and UNEP) issued a report on the health impacts that might occur in conjunction with climate change, titled Climate Change and Human Health — Risks and Responses ; in 2005, WHO also issued Using Climate to Predict Infectious Disease Epidemics about the impacts that climate change might have on infectious diseases. As stated in those documents, it is difficult to accurately predict the impacts of climate change on health. Infectious diseases and other health damage occur because of complexly interrelated factors including sanitary conditions, state of nutrition, host immunity, pathogenicity of pathogens, and routes of transmission. Therefore normally it is hard to predict how climate change that is only one of the parameters impacts will be manifested.

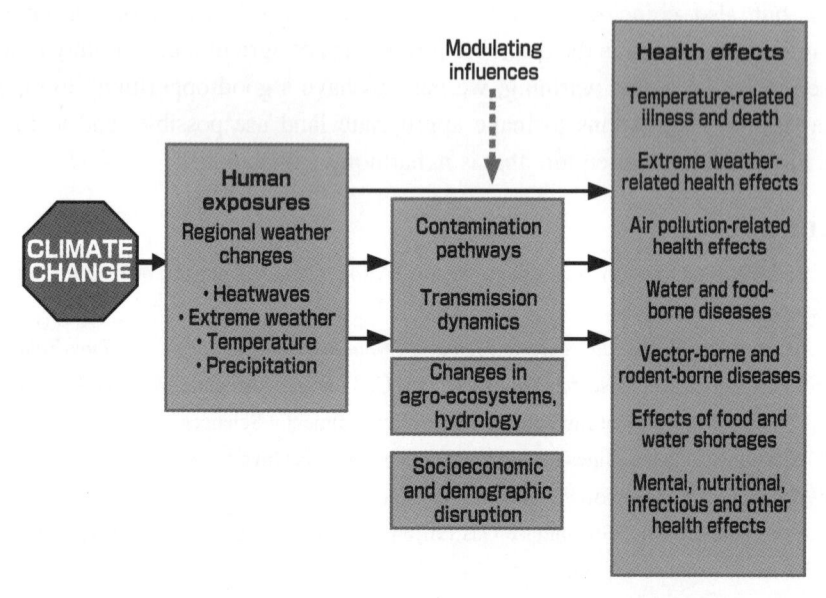

Fig. 1 Pathways by which climate change affects human health (modified from reference)

Health Impacts, Mainly Infectious Diseases, Due to Climate Change

Climate Change and Human Health − Risks and Responses summarizes the health impacts of climate change as shown in the diagram (Fig. 1).

Conceivable first of all are direct health impacts that occur as a direct result of climate change, such as the harm caused by heat waves and extreme weather events. Next is indirect harm such as changes in the patterns of infectious disease epidemics caused by, for example, environmental changes due to changes in temperature and precipitation and changes in transmission routes. A third possibility is health damage that occurs in a more indirect manner, such as impacts on agriculture and depletion of water resources owing to changes in temperature and precipitation, and impacts on the socioeconomic system. Regarding the health impacts of climate change, direct harm such as heat waves and increased infectious diseases are often addressed, but it is possible that the third mechanism will have the largest impacts over the long term.

There are also a number of conceivable mechanisms by which climate change affects infectious diseases. Some of these are: (1) The occurrence of infectious diseases due to extreme weather events, (2) increase in infectious diseases from water and food because of water and food shortages, (3) increase in infectious diseases whose vectors are animals or insects, in conjunction with increases or changes in the distributions of vectors, such as mosquitoes, that transmit infectious diseases, (4) increase in infectious diseases owing to ocean changes such as rises in sea level and in seawater temperature, (5) changes in the seasonality of infectious diseases that are seasonal, such as influenza.

Many infectious diseases are given as those which might possibly be affected by climate

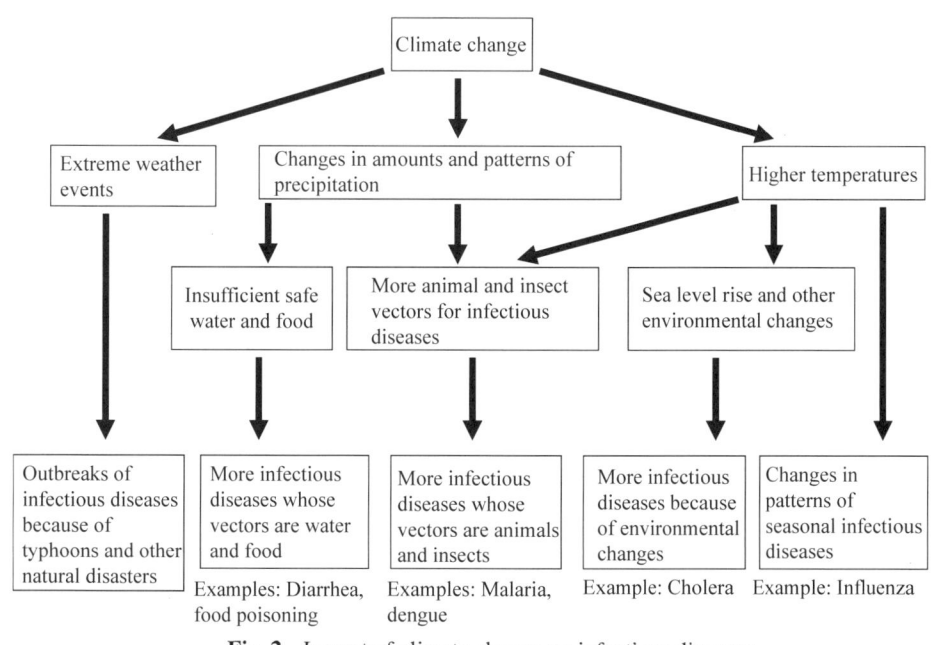

Fig. 2 Impact of climate change on infectious diseases

change, but only a limited number of these have been linked on scientific grounds to climate change. One of these diseases is dengue fever, for which it is thought that the expanded habitat of the vector mosquito due to global warming might lead to a larger geographical range for the disease. Cholera is always present in brackish waters, and it has been proved that cholera is more active when seawater becomes warmer. It is known that larger populations of the mosquito vector for Rift Valley fever, a serious infectious disease found in Africa, directly increase the number of human victims. Influenza and other seasonal infectious diseases might be considerably affected by climate change, but it is unknown what kind of impact there will actually be. It is also hard to predict accurately how many other infectious diseases will be affected by climate change.

Chapter 36

The IPCC Now : Walking the Fine Line between Neutrality and Policy Prescriptiveness

Anne McDonald

Twenty years have passed since the establishment of the Inter-governmental Panel for Climate Change (IPCC). Set up by the World Meteorological Organization (WMO) and the United Nations Environment Programme (UNEP), the IPCC was formed as a policy neutral scientific body with a mandate to provide decision-makers with scientific technical and socio-economic information about climate change *.

This paper is not concerned with an analysis of the actual findings of the IPCC reports, but rather will look at how the role(s) of the IPCC have evolved, specifically IPCC's role as a major source of information for global climate change related negotiations and as a voice in influencing climate change related policies both globally and regionally.

Of central interest is the role of IPCC's most recent report, the 4th Assessment Report, in Post-Kyoto negotiations. With increased scientific certainty of climate science, it may be argued that the voice of IPCC has incrementally strengthened; the co-awarding of the 2007 Nobel Peace Prize to IPCC attests to this.

Along with this elevated recognition and status, however, comes the question of whether the IPCC has been able maintain its policy neutrality or has it crossed the line over to being a policy prescriptive negotiating force.

While investigating the fine line between scientific neutrality and policy prescriptive inclinations, this paper will attempt to identify the potential roles for IPCC's forthcoming 5th Assessment Report, including roles to be played by Japan within the IPCC.

* IPCC mandate is "to assess scientific, technical and socio-economic research relevant to understanding the risk of human-induced climate change, its observed and projected impacts, adaptation and mitigation options available to policy makers".

Chapter 37

Climate Change Impacts, Adaptation, and Mitigation Measures : Findings of the Synthesis Report

Hideo Harasawa

Introduction

2007 was an important year for considering how to deal with global warming. The IPCC Fourth Assessment Report's Working Group I Report (The Physical Science Basis), released on February 2, 2007, showed scientifically that global warming is evident from the global temperature upswing and other observations, and that it is quite likely the cause is carbon dioxide and other greenhouse gases emitted by human activities. The Working Group III Report (Mitigation of Climate Change) released in April showed that holding the extent of global warming down to about 2℃ requires greenhouse gas emissions to peak in the next 10 – 20 years, and a cut of at least 50% in 2050, and that cutting emissions is possible by mobilizing current reduction technologies and by using economic incentives such as pricing carbon.

In November, at the 27th Session of the IPCC (November 12–17, 2007, Valencia, Spain), the Synthesis Report summarizing the three working group reports was adopted. UN Secretary–General Ban Ki–moon also attended, a press conference was held, and the report was released worldwide. This report was used at COP13, which was held on Bali, as basic material for discussion on the framework for 2013 and beyond. In Japan, the government announced the Strategy for an Environmental Nation in the 21st Century, under which Japan will halve its greenhouse gas emissions in 2050, and also announced Blue Planet 50, showing the world Japan's basic approach for its long–term strategy to combat global warming.

On October 12, before the IPCC session, the decision was made to jointly award the IPCC and former US Vice President Al Gore the Nobel Peace Prize. The reason given was that they had accumulated and disseminated scientific knowledge on anthropogenic climate change and laid the foundation for addressing climate change. Global warming has now developed into a problem so serious as to jeopardize world peace. All countries, both developed and developing, must draw on the IPCC's global warming prescription. While keeping an eye fixed on the long-term future, they must in the short term attain their reduction commitments for the Kyoto Protocol's First Commitment Period, and then they must develop plans for the more rigorous emission reductions to come, and build truly low-carbon societies.

Below I describe the IPCC Fourth Assessment Report (especially the Synthesis Report), which is especially important as the scientific basis for considering how to deal with global warming.

IPCC fourth assessment report and the synthesis report

The Fourth Assessment Report comprises reports by the three working groups and a comprehensive Synthesis Report prepared on the basis of those three reports. Based on the scientific knowledge gained after the Third Assessment Report (2001) was released, the Synthesis Report summarizes in a crosscutting and comprehensive manner the scientific knowledge brought together in the working group reports. This knowledge includes the phenomenon of climate change, its causes, predictions, impacts, adaptation, and mitigation. The Synthesis Report consists of the full text and the Summary for Policymakers (SPM). The Synthesis Report's SPM was prepared to provide the world's people and especially policymakers and politicians with the latest scientific knowledge in a brief form. The Synthesis Report is based on the full text and SPM of each working group report; it uses many tables and graphs, and is easy to read and understand.

The Synthesis Report encompasses the following six topics.
1) Observed changes in climate and their effects
2) Causes of change
3) Projected climate change and its impacts
4) Adaptation and mitigation options
5) The long-term perspective
6) Robust findings, key uncertainties

The sixth is contained only in the full text. The full text and SPM are available in the originals and Japanese translation (IPCC [2007], Ministry of Education, Culture, Sports, Science and Technology [2007]).

Topic 1. Observed changes in climate and their effects

This section presents the observed climate changes and their impacts on humanity and natural systems.
- Warming of the climate system is unequivocal, as is now evident from observations of increases in global average air and ocean temperatures, widespread melting of snow and ice, and rising global average sea level.
- Many natural systems are being affected by regional climate changes.

Topic 2. Causes of change

Presents the observed causes of change.
- Global GHG concentrations are much higher than the level of pre-industrial times due to human activities.
- Most of the observed increase in global average temperatures since the mid-20th c

entury is very likely due to the increase in anthropogenic GHG concentrations.

Topic 3. Projected climate change and its impacts

Presents short- and long-term climate change and its impacts based on various suppositions about the future (emission scenarios).

- It is projected that if current policies are continued, global GHG emissions will continue to grow over the next few decades, and global warming in the 21st century will exceed that observed in the 20th century.
- It is predicted that many changes will be brought about in Earth's climate system. Predictions include impacts by sector and when they appear, impacts anticipated for certain regions, and extreme phenomena (such as extreme weather events).

Topic 4. Adaptation and mitigation options

Arresting global warming requires reducing emissions of greenhouse gases, which are the cause. Ways to reduce greenhouse gas emissions are called mitigation. The impacts of global warming are already evident around the world, and it is predicted that if warming continues, its impacts will appear in various sectors, and in both the developed and developing countries. This topic discusses adaptation and mitigation, and describes the relationship with sustainable development on global and regional levels.

- Reducing vulnerability to climate change necessitates stronger adaptation strategies than at present. The report gives examples of specific adaptive measures for different sectors.
- Implementing appropriate mitigation measures could offset and reduce the increase in global greenhouse gas emissions over the next several decades.
- In the report's estimation, the UN Framework Convention on Climate Change (UNFCCC) and the Kyoto Protocol have built the foundation for future efforts at mitigation as an international framework for encouraging mitigation

Topic 5. The long-term perspective

As a long-term perspective, and especially in accordance with the ultimate goals and provisions of the UNFCCC, this section presents the scientific and socioeconomic aspects of adaptation and mitigation in connection with sustainable development.

- The following five "reasons for concern" in relation to climate change, which the Third Assessment Report identified, are even stronger.
    1. Increased risks to unique and threatened systems, such as polar and high-mountain communities and ecosystems.
    2. Increased risks of extreme weather events such as droughts, heat waves, and floods.
    3. Major impacts and vulnerabilities affect those who are regionally and socially in weak positions.
    4. The benefits of global warming will peak at lower temperatures. As warming proceeds, the damage will worsen, and the costs of global warming will increase with time.
    5. There will be increased risks of major change, such as accelerated sea level rise and ice sheet loss.
- It is possible that neither adaptation nor mitigation measures alone will be enough, but

they can complement each other and together can considerably lower the risks of climate change.
- Stabilization of greenhouse gas concentrations is possible with existing technologies and those that will become available in coming decades. The key is the mitigation effort and investment made in the next 20-30 years.

Topic 6 appears only in the full text. It is a discussion on robust scientific findings and uncertainties. This topic was deleted from the SPM.

The significance of the fourth assessment report : Center for global environmental research, 2007

Preparation of the Fourth Assessment Report concluded with the release of the Synthesis Report; however, future IPCC sessions are planned to evaluate the Fourth Assessment Report and IPCC activities. The significance of the Fourth Assessment Report will probably be discussed, but the significance at this point in time can be summed up as follows.

1) Warming of the climate system is unequivocal.

Climate change observations and elucidation of phenomena have made progress. Warming of the climate system is assessed as very likely, and it is nearly conclusive that the causes of global warming are greenhouse gas emissions and other human activities. For such reasons the certainty of scientific knowledge on climate change improved substantially.

2) The impacts of global warming are now apparent.

On all continents and in almost all oceans, it is now clear that the natural environment, such as snow, ice, and ecosystems, as well as human activities, are affected.

3) Global warming has impacts on various sectors and regions.

It is predicted that at the end of the 21st century the average global temperature will be 1.1 -5.8 ℃ higher, and sea level will be 18-59 cm higher, than in the 1990s, with impacts appearing in various sectors and regions. A rise of 2-3 ℃ in temperature over that of the 1980s and 1990s would temporarily bring favorable effects (for example, warming in cold regions would make grain cultivation possible), but harmful impacts would prevail with greater temperature increases.

4) Climate change must be addressed quickly.

To arrest global warming, it is necessary to achieve a downward curve in greenhouse gas emissions during the next 20 to 30 years so that emissions are substantially reduced in 2050. The report identified the relationships between long-term stabilized concentrations and remedial measures, which will aid consideration of the post-Kyoto framework.

5) Mitigation is cheaper than the damage.

In the way of mitigation measures, greenhouse gas emissions can be sufficiently reduced with current technologies, economic measures, and changes in lifestyle and consumption patterns. And if one takes the cobenefits into consideration, the economic costs are lower than the cost of global warming damage.

6) Both mitigation and adaptation are necessary.

Mitigation to prevent global warming and adaptation to alleviate the impacts of global warming are both needed. Combining them well makes it possible to reduce the risks of global warming with limited funds. But there are still various constraints on implementing both.

Future developments and Japan's contribution

The Fourth Assessment Report has been completed, but already activities for the Fifth Assessment Report have begun. The fifth report is expected to be released around 2013, and because the writing will take about three years, various activities will probably start this year. It is important to pursue work in the following areas for Japan's contribution to the fifth report in terms of research.

- Release of peer reviewed English-language papers: Because the IPCC uses peer reviewed papers as an information source, it is important as always to release papers. Japanese-language reviewed papers are also used for the assessment if they have English-language abstracts. It is also effective to send papers to the IPCC and to the report's writers and other well-known researchers.
- Participation as writers: Now that the Fourth Assessment Report is finished, over the next one or two years there will be scoping meetings and other meetings to set up next system (election of chairperson and bureau). These meetings will discuss the issues to be addressed by the next report and decide the draft table of contents. After this, the writers are chosen. Thirty Japanese researchers contributed to the Fourth Assessment Report, and we hope that even more will contribute as writers to the fifth report.
- Review of Japan's global warming research: It would also be effective to conduct a review of papers on global warming to summarize the achievements and findings of Japanese research and publish the review as an English-language report or book.
- Active participation in IPCC workshops and other meetings: From now on the IPCC is expected to hold frequent workshops on various issues. It is important to actively participate in such workshops and present Japanese research.
- Support for global warming research in Asian developing countries: APN (Asia-Pacific Network for Global Change Research) and other organizations lend support (such as support for research on impacts in developing countries), but currently there is too little support by Japanese researchers, which makes pursuing research extremely difficult. It is important for Japan to provide cooperation and support for global warming research conducted by Asia-Pacific countries.

References

1) Center for Global Environmental Research (2007).
   Highlights of the IPCC Fourth Assessment Report, 12 pp.
   http://www-cger.nies.go.jp/cger-j/pub/pamph/pamph_index-j.html#ipcc>
2) Ministry of Education, Culture, Sports, Science and Technology; Ministry of Economy, Trade and

Industry, Meteorology Agency; Ministry of the Environment (2007). *On the Release of the* ss kit.

3) IPCC (2007) Summary for Policymakers of the Synthesis Report of the IPCC Fourth Assessment Report. 23 pp.

Part VI
Food Safety and Preventive Medicine

38	A Message from the Symposium Organizer ·············· 169 *Tadayoshi Shiba*	
39	Food Safety Commission : The First Five Years and Coming Challenges ····· 171 *Takeshi Mikami*	
40	Current Dietary Lifestyle and Its Challenges : Linking Health Maintenance, Good Taste, and Safety 1. The Role of Registered Dietitians in Dietary Life ············· 175 *Masaki Taga*	
41	Current Dietary Lifestyle and Its Challenges : Linking Health Maintenance, Good Taste, and Safety 2. Kitasato Yakumo Beef Cattle ······················· 177 *Kumiko Asahi*	
42	Current Dietary Lifestyle and Its Challenges : Linking Health Maintenance, Good Taste, and Safety 3. Using Dietary Fiber in Foods, and Preventive Medicine ········· 179 *Masashi Omura*	
43	Functions and Safety of Fishery Products ················ 181 *Hisao Kamiya*	

44   Lipid Peroxides and Disease · 185
     Yasuhito Nakagawa

45   Salmonella and Campylobacter Foodborne Disease : An Farm Level Perspective
     · 189
     Masayuki Nakamura

46   Problems Involving Arsenic :
     Induced Health Effects in People Who Consume Much Seaweed · · · · · · · · · · · · 195
     Hiroshi Yamauchi

47   The Role of the Center for Genetic Studies of Integrated Biological Functions
     in Agromedicine :
     The Kind of Laboratory Animal Facility Needed for Future Medical Research
     · 200
     Nobutaka Shinohara

48   Possibilities of Probiotics as a Bridge for Agriculture – Medicine Collaboration
     · 203
     Takao Mukai

49   Possibilities and Limitations of Functional Foods · 208
     Keizo Arihara

50   Current State of the Kitasato University Agromedicine Concept · · · · · · · · · · · · 213
     Katsu Minami

Chapter 38

A Message from the Symposium Organizer

Tadayoshi Shiba

I would like to offer a few opening remarks on behalf of the sponsor of this Sixth Kitasato University Agromedicine Symposium.

There is no way to overemphasize the need for agromedical science and education in fields such as disease prevention, health improvement, safe food, conservation agriculture, and healing agriculture; in other words, for the health and safety of people living in the 21st century.

This agromedicine symposium presented by instructors at Kitasato University, who are explorers of the life sciences, is titled "Food Safety and Preventive Medicine".

The situation surrounding dietary lifestyle and preventive medicine has undergone sizable changes. These include globalization of food, the appearance of new hazards such as *Escherichia coli* O157 and misfolded prion proteins, the development of new technologies such as genetic recombination, and poisoning incidents typified by Chinese-manufactured foods. It has been some time since scientific information on food safety and preventive medicine started making its way deeply into our everyday lives.

The consequence is a trend that obliges us to have a constant consciousness of science at the dinner table. The anti-oxidant effect of red wine's polyphenols is emphasized, while nothing is said about the problem of what increased alcohol consumption does to health. You are conscious of avian flu just eating one stick of grilled chicken. And when you look at the packaging of gyoza, you wonder where they were made.

Another trend now is to emphasize the efficacy and impacts of the constituent parts of a food, while saying nothing about what happens when they are put together. One type of knowledge is isolated from another.

Because foods are an amalgam of many constituents including carbohydrates, fats, proteins, minerals, and vitamins, they have both upsides and downsides. Since ancient times we have had great concern for taste and preservation methods to keep a delicious, well-balanced diet. Everyone knew that when foods have no additives, they spoil and increase the risk of food poisoning.

Through our long human history, people have gained the wisdom to eat a balanced diet of various foods. It seems that modern food and preventive medicine with their underpinning of science and technology are making us forget our human history, practices, and common sense.

The latest food safety and preventive medicine are sustained by drawing on the latest science and technology. My hope is that this symposium will consider, from the standpoints of agricultural science and medical science, how this food safety and preventive medicine can collaborate.

In closing, I would like to express my sincere gratitude to the speakers who gladly agreed to address this symposium.

Chapter 39

Food Safety Commission : The First Five Years and Coming Challenges

Takeshi Mikami

Introduction

The Food Safety Commission was established in July 2003 as an "organization to objectively, neutrally, and impartially assess food safety based on science". This July it was five years old. Today I will review the commission's initiatives over the first five years and outline its future challenges.

Creation and role of the food safety commission

1) Contexts of the commission's establishment

The circumstances surrounding our dietary lives have undergone considerable changes, including the advancing geographical broadening and internationalization of food distribution, the appearance of new hazards such as O157 and abnormal prions, the development of genetic recombination and other new technologies, and improvements in analysis techniques.

Meanwhile, a new approach to food safety is a risk analysis method that has become generally adopted internationally. In this approach, all foods should be scientifically assessed and controlled under the assumption that they present risks. Risk analysis methods comprise the three elements of risk assessment, risk management, and risk communication. Risk assessment involves scientifically assessing the health impacts of ingesting hazards in food. Risk management is carried out, based on risk assessment results, for purposes such as deciding pesticide residue and usage of standards, and pesticide regulation, while taking into consideration of cost-effectiveness, technological possibilities, and popular sentiment. There is also supposed to be risk communication, involving the exchange of views and information, through risk assessment and risk management with involved parties including consumers and food businesses.

2) Creation and role of the food safety commission

Under these circumstances described above, Japan's first case of BSE surfaced in 2001, which elicited scathing criticism of the government's response from the citizens. Based on a sense of remorse over food safety administration until that time, in July 2003, the Food Safety

Basic Law entered into force and the Food Safety Commission was established under the Cabinet Office.

The purpose is to incorporate risk analysis methods into food safety administration and to perform food safety-related risk assessment independently of the Ministry of Agriculture, Forestry and Fisheries (MAFF) and the Ministry of Health, Labour and Welfare (MHLW), while according the highest priority to protecting citizens' health.

In addition to risk assessment, the commission carries out risk communication, and in the event of a mass food poisoning incident, it is responsible for gathering information, informing the citizenry, and taking other emergency actions to respond.

The Food Safety Commission is composed of seven commissioners, and has 14 expert committees. The "Planning", "Emergency Response", and "Risk Communication" committees deal with crosscutting issues, and the remaining 11 expert committees study and discuss risk assessment for individual hazards such as additives, pesticides, veterinary medicines, microorganisms/viruses, prions, and genetically modified foods.

Initiative and achievements over the first five years

1) Initiatives over the first five years

Risk Assessment

The Food Safety Commission's most important role is risk assessment. In response to requests from the MHLW and the MAFF, which are risk management organizations, the commission conducted over 600 risk assessments in this five-year period. The commission also conducts "self-initiated assessments", which are risk assessments on subjects chosen by the commission at its own discretion.

Assessments on BSE are important assessments that test the true value of the commission. The commission collected information and data widely in Japan and other countries, conducted careful scientific discussions, and performed risk assessments including "Medium-Term Summary of Measures to Address BSE", "A Review of Japan's Measures to Address BSE", and "A Risk Assessment of Beef and Other Products Originating in the United States and Canada."

In addition to BSE, the commission carries out risk assessments on *Campylobacter jejuni*, *C. coli*, and other microorganisms that cause food poisoning, and on drug-resistant bacteria that are selected for by the use of antibiotics on livestock and other animals.

Furthermore, in conjunction with the introduction of the "positive list system" in May 2006, the commission is performing a series of assessments on 758 substances including pesticides and veterinary medicines.

Risk Communication

The Food Safety Commission endeavors to use every opportunity and communication tool to carry out risk communication, including opinion exchange forums held around the nation and by seeking opinions and information on such things as risk assessment results, as well as by releasing information on the web and providing information through various media including printed materials and DVDs. In our first five years, we held 324 opinion exchange forums, and sought opinions and information for 353 times.

Starting in 2006, the commission has been cooperating with municipalities by holding regional leader training courses to foster people who will actively facilitate risk communication in outlying regions.

What is more, in order to avert social turmoil or excessive concerns among the populace, the commission takes advantage of its nature as an organization conducting scientific risk assessments to release media statements from the chair of the commission on matters vitally related to health damage from eating food, such as with avian flu.

2) Five years' achievements

Since July 2003, the Food Safety Commission has performed risk assessments objectively, neutrally, and impartially as an independent organization conducting science-based deliberations, while maintaining a relationship with risk management organizations in a collaborative but restraint manner. Our five years' effort for the initiatives seems to be paying off, because the idea of risk analysis has found a home among those who are involved in food safety.

In addition, owing to the creation of the commission, works such as in laying the foundation of risk communication, to secure as in securing transparency in the risk assessment process by releasing the deliberation process, and regularly using tools such as opinion exchange forums and seeking opinions and information. We have also strengthened our provision of food safety information to the public.

Future challenges facing the commission

Upon the five-year milestone of the Food Safety Commission, we must summarize our five years of accomplishments and review our operations. The need for the reform of the Food Safety Commission has also been pointed out in discussions by the Consumer Affairs Administration Promotion Council and other venues during the creation of a Consumer Affairs Agency.

Bearing in mind the attention and expectation of the parties concerned, the commission intends to work on the challenges detailed below to further strengthen its functions and role.

1) Risk assessment

In the area of risk assessment, the commission has responded to the increasing number of cases to be discussed, and has made improvements including reviews of its deliberation system and methods among others. Nevertheless, many cases remain to be assessed. Many more assessments are expected due to the factors such as the positive list system and the development of new technologies. Therefore, it is necessary to conduct risk assessment deliberations more effectively and efficiently.

Some reputations on the self-initiated assessments are that "there are only few cases" and "the commission should assess more generally concerned cases by consumers". To satisfy those views, the selection process for the items of self-initiated assessment needs to be re-examined. Additionally, tasking self-initiated assessments will require furthermore data/information correction. The data analysis system also will need to be reinforced.

2) Risk communication

Some reputations on risk communication are that "assessment results are hard to understand" and "the mutual flow of information and opinions is not assured". In addition, municipalities have various degrees of commitment to risk communication initiatives, and efforts to foster people to sustain risk communication in outlying regions have just begun.

In view of this situation, it is necessary to make changes. Preparing materials on risk communication in more easily understandable manner, reforming opinion exchange forums and other gatherings for more participation, and improving the way the opinions and information are sought. It will also be necessary to support the autonomous initiatives of the municipalities, and to facilitate the commission's collaboration with them.

3) Information provision

In the area of information provision, it is doubtful if there is broad and sufficient understanding among the public about the risk analysis approach introduced by the Food Safety Basic Law. Same goes to the roles, activities, and other aspects of the Food Safety Commission. At a time when food-related problems are arising, there is a need for information provision that will moderate the public concerns about food safety.

This creates a need to make active use of diverse media and means for the provision of more easily understandable about risk assessment framework and the role of the Food Safety Commission. Another challenge is how to make the timely release of chairperson's media statements and other information when potentially alarming food safety problems arise for the public.

4) Internationalization of commission activities and international linkages

In the age of food globalization, it is important that not only Japan actively participates in creating international standards, but also that consistency of domestic Japanese standards with international standards. Due to having introduced the changes such as the positive list system, there is a sharp increase in the number of cases demanding attention to international trends in relation to risk assessment.

In view of these circumstances, it is essential to quickly obtain information on risk assessment results and assessment methods in other countries, and to cooperate in the work of international risk assessment. The Food Safety Commission will need to strive on strengthening collaboration with international agencies and organizations of other countries. In particular, the commission intends to bolster its collaboration with the European Food Safety Authority (EFSA) such as through the exchange and sharing of everyday information.

Conclusion

As of the 248th meeting on July 24, 2008, the Food Safety Commission considers its way of improvement. In view of the necessary steps ahead, the commission plans to hold discussions and produce a final summary of improvement measures before the end of FY2008.

Chapter 40

Current Dietary Lifestyle and Its Challenges: Linking Health Maintenance, Good Taste, and Safety

1. The Role of Registered Dietitians in Dietary Life

Masaki Taga

Food is essential for maintaining human life. Since ancient times we have had the dietary habit of preparing ingredients and eating food together. Everyone wants to have sound dietary habits and lead a healthy life. In other words, food has roles that are important both as the foundation of daily life and medically, such as by maintaining health, improving the quality of life, and preventing and curing illness. At the same time, many issues have arisen with respect to the safety and functionality of food, and concerns about ingredients and foodstuffs are now much greater than in the past. Dietary lifestyle is said to include: (1) maintaining our physical health, (2) maintaining our mental health, (3) maintaining social health, (4) passing on culinary culture to future generations, and (5) education. By means of eating, we have used these functions to maintain life.

But in postwar Japan, with its scientific and technological development and economic growth, our dietary lifestyle has transitioned into gourmet pursuits, instant food, and "healthful" eating. For this reason, lifestyle diseases such as diabetes and metabolic syndrome are now of great significance, while the linkage of health maintenance, good taste, and safety is important to the dietary lifestyle of Japan because of its high median age.

More and more people now subscribe to the thinking that to prevent lifestyle diseases, even if one falls victim to metabolic syndrome or diabetes, we should to the greatest extent possible avoid medicine and cure and prevent illness by managing our own health with food and exercise. Owing to this state of the modern dietary lifestyle, there are high expectations for the expertise of registered dietitians, whose job is to prevent and cure illness with food. Registered dietitians appeared with the partial revision of the Dietitian Law in 1962, wherein they were defined as people who provide the complex, difficult nutritional guidance that ordinary dietitians cannot do. A 2000 revision of the Dietitian Law defined registered dietitians thus: "The title of registered dietitian is conferred when licensing by the MHLW to people whose occupation is: to provide nutritional guidance necessary for the medical treatment of the sick and injured; to provide nutritional guidance for the maintenance and improvement of health requiring advanced specialized knowledge and techniques corresponding to the physical condition, nutritional condition, and other attributes of individuals; and to manage the preparation of meals requiring special consideration for the physical condition, nutritional condition, type of use, and other attributes of people using facilities which provide meals on a continuing basis for large numbers of specific people, as well as to provide guidance and other assistance needed to improve nutrition in such facilities."

Dietetics, which is the specialized and academic foundation of registered dietitians, is made up of three elements: the human body and health; food and nutrients; and food-related behavior and the social environment. In other words, dietetics is an academic system comprising the fields of medical science, agricultural science, and home economics. Registered dietitians are involved in research and education on the functionality of the various constituents of foodstuffs, education on primary prevention by improving dietary habits (nutritional education), secondary prevention (such as specific medical checkups and health guidance for metabolic syndrome), and welfare as tertiary prevention (facilities for the elderly). Registered dietitians are also involved in education for knowledge and techniques for bedside nutritional management in the treatment of disease (for members of nutritional support teams in team medical treatment). Here we describe two examples of our efforts to link agriculture and medicine in the context of Current Dietary Lifestyle and Its Challenges: Linking Health Maintenance, Good Taste, and Safety.

Chapter 41

Current Dietary Lifestyle and Its Challenges: Linking Health Maintenance, Good Taste, and Safety

2. Kitasato Yakumo Beef Cattle

Kumiko Asahi

Nutritional state of the citizens : Results of national health and nutrition survey

In recent years, because of the increase in lifestyle diseases, obesity, and the like, attention has been focused on nutritional problems. Further, in April 2008 the government started providing specific medical checkups and health guidance for all citizens 40 years or older, and nutrition for health improvement is considered important. According to the results of the 2006 National Health and Nutrition Survey, there were 5 million more people with a lifestyle disease (diabetes) than four years before, and among people aged 40 to 74, one in two men and one in five women were strongly suspected of having metabolic syndrome or getting it in the future. There were more obese men in all age groups than 20 and 10 years before. In particular, 30% of men aged 30 to 60 were obese.

In terms of nutrient intake, there was a gradual trend toward ingesting less energy. However, there was also a gradual increase (about 20% of men and 30% of women 20 years and older) in the number of people who ate less carbohydrates and derived 30% or more of their energy from fat. With regard to dietary habits, the higher percentage of people who skip breakfast and the trend toward later dinner times suggest the need for improved dietary habits.

Nutritional characteristics of Yakumo beef

Kitasato Yakumo cattle yield safe, worry-free Healthy Beef because they are 100% raised on farm-grown pasture on which no chemical fertilizers have been used at all. Nutritionally the beef is superior to that of conventionally fattened cattle and is characterized by constituents having more tertiary functions such as anticancer action and arteriosclerosis prevention.

In particular, the fatty acid composition is lower than that of marbled beef, with a ratio of $n-6$ to $n-3$ fatty acids around the appropriate value of 4. Kitasato Yakumo Healthy Beef has a high concentration of conjugated linoleic acid (for which various physiological functions such as obesity prevention, improved fat metabolism, and improved insulin resistance are reported), at about twice that of conventional beef. What is more, Kitasato Yakumo Healthy Beef contains high levels of iron (which when deficient will lead to anemia), zinc (which when insufficient leads to gustatory disorders and lowered immunity), and magnesium (which plays an important role in energy metabolism). Therefore, it is a food nutritionally appropriate for preventing lifestyle diseases.

Initiatives by the Kitasato junior college of health and hygienic sciences' applied clinical dietetics department

This academic year the college launched a project to develop processed foods and recipes for hospital meals using Kitasato Yakumo Healthy Beef. We tackled the development of cuisine that takes advantage of the meat's nutritional characteristics and flavor as an integral part of education. In a sensory test involving students, many subjects said that although Healthy Beef was inferior to Aussie Beef in terms of tenderness and smell, its appearance (color) and taste was superior. We offered curry with our beef to patrons at a lunch during the college PPA meeting, and only 17% of them said the beef was tough. For that reason we have classes in which students think about cooking methods which take advantage of the characteristic firmness and flavor, and produce delicious cuisine.

Henceforth we intend to take advantage of the meat's nutritive and sensory characteristics, and from the perspective of preventing lifestyle diseases, practice the development of processed foods and hospital meals using Yakumo Healthy Beef.

Chapter 42

Current Dietary Lifestyle and Its Challenges: Linking Health Maintenance, Good Taste, and Safety

3. Using Dietary Fiber in Foods, and Preventive Medicine

Masashi Omura

Introduction

Dietary fiber used be considered worthless in terms of nutrition, but now it is spotlighted as the "sixth nutrient" along with proteins, fats, sugars, vitamins, and minerals, and is seen as useful in preventing colon cancer and lifestyle diseases.

Although dietary fiber has long existed, it has a short research history, and it is still a poorly understood field. However, preventive medicine has big expectations for the direct and indirect bioactivity of dietary fiber.

Here I report the latest findings on the definition and classification of dietary fiber, and the relationship of its bioactivity to health. Additionally, I report on the possibilities and food applications of resistant starch, which has an especially unusual function.

Dietary fiber classification and definition

The types of dietary fiber include soluble, insoluble, plant, animal, bacterial, natural, artificial, and chemically modified. In Japan, the definition of dietary fiber is "a general term for indigestible constituents of food that are not digested by human digestive enzymes".

The Japanese Association for Dietary Fiber Research proposes the word "lumiconide" as a blanket term for substances, including dietary fiber, which are indigestible, unabsorbable, and have some kind of physical action in the digestive tract. Their definition is "a food constituent that is difficult to digest or absorb in the human small intestine, and manifests a physiological effect via the digestive tract that is useful in maintaining health".

Dietary fiber bioactivity

From the experimental, clinical, and epidemiological verification results to date, it is known that dietary fiber manifests a variety of physiological effects in the process of passing through the digestive tract. Some representative physiological effects are that fiber effectively regulates the intestinal environment, facilitates bowel movements, and prevents colon cancer; it is also effective for preventing lifestyle diseases such as diabetes, hyperlipidemia, hypertension, and obesity.

In addition to being generally low energy, dietary fiber is fermented by intestinal bacteria in the colon resulting in the formation of short-chain fatty acids. These short-chain fatty acids facilitate the proliferation of colonic epithelial cells and suppress the synthesis of cholesterol, which offers promise of a new effect that was previously unknown.

Application of resistant starch in foods

Resistant starch is defined as "a general term for starch and partially water-soluble starch that is not digested or absorbed in the small intestine of a healthy person". There are four types of resistant starch categorized from RS1 to RS4.

A feature of resistant starches is that they combine the properties of both soluble and insoluble dietary fiber, and are known to have various beneficial effects including improving bowel movements, mitigating colon disorders, lowering blood cholesterol, improving blood sugar level, improving mineral utilization, and producing short-chain fatty acids from fermentation by colonic bacteria.

Resistant starch is also excellent for food processing. For example, it can be added to starchy foods such as bread, noodles, pizza, and other staples. An advantage is that it can be added in large quantities without impairing texture, flavor, or aroma. Thus, one can continuously consume it naturally from daily meals. Presumably using resistant starch effectively makes it possible through ordinary meals to improve one's intestinal environment and prevent lifestyle diseases.

Chapter 43

Functions and Safety of Fishery Products

Hisao Kamiya

Fisheries industry target species

The oceans and the rest of the hydrosphere are populated by a vast diversity said to comprise 500,000 or even 5 million species from plankton to cetaceans. Of those, the fisheries industry is concerned with perhaps a few percent, so hardly any marine organisms are used.

The fishery industry's main mode of production is the exploitation of wildlife living in the natural environment. Breeding and aquaculture in the sea or inland waters amounts to the same thing. In this respect, the fisheries industry mode of production differs from that of terrestrial agriculture and livestock husbandry, where people raise and manage animals that have been selectively bred for a long time. Further, the fish and shellfish exploited by the fisheries industry are subject to a phenomenon not seen in species used for land-based food production in that some species include tissues that are highly toxic, while other species become toxic due to changes in their habitats, which are natural phenomena.

Functions of fishery products

Epidemiological studies indicate that human disease incidence rates are lower and lifetimes are longer in regions where people eat much rice and fishery products, and fishery products have an established reputation as healthful. In fact, seafoods are rich in high-quality proteins, lipids, vitamins, minerals, and more, and are excellent as a source of nutrients. It is known that the n-3 OK polyunsaturated fatty acids such as icosapentaenoic acid (IPAOK), which are plentiful in blue-backed fish such as sardines and mackerel, have effects such as lowering levels of blood cholesterol and neutral fats. Moreover, although there is variation in the methods and levels of scientific assessment underpinning the claims, it is reported that seafood constituents have many kinds of bioactivity including antibacterial activity, hypotensive activity, antioxidative activity, immunomoduratory activity, and antitumor activity.

IPA OK is the only functional active substance exhibited by seafood that has been identified and developed into pharmaceuticals, but there have been vigorous efforts in the development of functional foods in the quest for higher added value. As a result, there are many seafood-derived substances used as constituents in "foods for specified health uses", which are authorized by the Minister of Health, Labor and Welfare, and substances used as the ingredients in so-called "health foods", which require no particular permission or authorization.

This year "specific medical checkups and health guidance" started with the aim of improving the situation in which many middle-aged and elderly Japanese are said to be at risk of developing lifestyle diseases. And on a daily basis advertisements feature words such as "metabolic", "health", "health foods", "supplements", and "natural foods". Although people understand the importance of improving energy-excessive dietary habits and eating balanced meals, they cannot easily change their meat-rich diets due to reasons such as price, convenience, and preferences. People think, "Well then, we'll switch to health foods and supplements"; but perhaps these are products of our era that were created by the desire of Modern Man to obtain "health" without any effort.

While health foods and related merchandise are said to have built an 800 billion-yen market, there are now concerns about problems of excessive expectations for their efficacy and about excessive use. Moreover, products whose ingredients and labeling are illegal, whose efficacy is suspect, and which have caused health damage and economic loss are confusing consumers who want health foods to be "worry-free". It is very attractive to seek added value in the "biological regulation function" of food, which is said to be its tertiary function; however, trustworthy claims about functionality need to be supported by medical and dietetic evidence, which is a high barrier to clear.

The safety of fishery products

We are fairly well versed in the kinds and ways of eating fishery products: there are no people who eat the liver or ovaries of Tetraodontidae species, we know what fish and shellfish caught on the coast are edible, which taste good and which do not, which are toxic and which parts of them are toxic, and we know to blanch seaweed in boiling water before eating. If fishery products are left sitting they will readily spoil and cause food poisoning, but there has been a sense of faith that fish and shellfish from near-shore waters are safe to eat.

To secure a source of protein in the postwar years, the development of new fisheries such as pelagic fisheries in distant waters and deep-sea trawling was promoted vigorously, which brought totally new species to our dinner tables along with Japanese-sounding names attached to them. Among the fish caught in distant waters and imported from foreign countries, people have found fish containing constituents whose safety cannot be confirmed, and those that closely resemble offshore species of Tetraodontidae whose flesh is highly toxic. With instances of food poisoning from South American fish resembling the Pacific rudderfish (Psenopsis anomala) and the green toadfish caught in Vietnamese waters, which resembles Takifugu spp., the Japanese are finding that their "common sense" gained through experience with fish from Japan's near-shore waters does not work with fish from other countries.

Recently a person who ate rocky porgy at a restaurant suffered ciguatera and filed suit against the chef under the Product Liability Law. The court said that other cases of ciguatera from the fish had occurred in Okinawa, rejected the claim that the defendant could not have had foreknowledge of its toxicity, ruled that the defendant could not avoid "the liability of the manufacturer in processing and providing poisonous fish", and recognized the plaintiff's claim for payment of damages. This court decision will likely become a precedent that restaurants that prepare and provide poisonous fish and shellfish have "product liability". Henceforth, people involved in the distribution and consumption of fishery products will need specialized knowledge on not only the characteristics of fishery products, but also on the safety of poisonous fish and the like.

Especially problematic with poisoning from natural toxins in fishery products are ciguatera and shellfish poisoning. Both types of food poisoning are caused by a natural phenomenon, the toxification of fish and shellfish by the blooms of toxic dinoflagellates. Because one can safely eat the same fish and shellfish from marine areas where toxic dinoflagellates are not present, there is great danger to importers and recreational fishers who lack knowledge and experience. We need to be aware that the presence of poisonous fish and shellfish previously limited to certain marine and geographical regions is now a direct danger that threatens the health of people around the world owing to globalized distribution of aquatic products. Shellfish that are especially popular worldwide sometimes cause fatal food poisoning, as with paralytic shellfish poisoning, and it is essential to monitor the toxicity of not only domestically produced fishery products, but also those imported.

To people who lived near the sea, fish and shellfish were a valuable, comparatively easy-to-obtain food. As one can see from the Tetraodontidae bones found among the fish bones in the shell mounds throughout Japan, people were also familiar with poisonous species. To safely enjoy the taste of fish and shellfish and to create our current healthful "fish-eating culture", there was no doubt much trial and error involving sacrifice. It is our responsibility to take the new knowledge about safety and functionality obtained with the latest science, incorporate it into this culture, and correctly communicate it to the next generation.

Securing fishery products, and marine life sciences

Marine ecosystems are built on a subtle balance among species that depend on that special environment called the ocean. It is the phytoplankton, microorganisms, and other autotrophic microorganisms and the primary producers that provide organic matter to marine organisms and sustain marine ecosystems. The food chain is composed of the zooplankton and other primary consumers (which are also secondary producers), as well as carnivorous and other high trophic level predators (which are also producers). The fisheries industry takes fish and shellfish that are at the top of the food chain. Therefore, it is possible that this balance will be disrupted not only by overfishing of species sought by the fisheries industry, but also by pollution of the marine environment that disturbs the amounts and distributions of plankton and of fish and shellfish lower on the food chain. It is also undeniable that overprotection of certain species affects the food chain. From a quite early time the fisheries industry was asked to make itself into a resource-protecting, environment-conserving industry that strives to peg catches to stocks. Ensuring that fishery products are sustainable means catching the surplus portion while leaving enough of a resource to reproduce itself. It is hard to imagine developing a fisheries industry that is compatible with resource protection while not having an understanding of marine ecosystems, such as the life stages of the target fish and shellfish from hatching to sexual maturity or its own relationship with many species and the marine environment.

Analyses of marine ecosystems have been carried out in fields such as population ecology, community ecology, and the environmental sciences.

Recently the most advanced chemical and biochemical technologies have been enthusiastically introduced in marine ecosystem analyses??? and research in this field has achieved a quantum leap. In the present lecture, I would like to focus on t he physiologically active substances that provide valuable information for elucidating on a molecular level the interactions between organisms of the same and different species. There are many instances in which research on the treatment and prevention of food poisoning by marine organism toxins, or on detection methods, has led to the development of new pharmaceuticals and biochemistry research reagents. There are unlimited possibilities in both basic and applied marine life science research that will be useful in securing fishery products and in their safe consumption. At Kitasato University as well, it is important to conduct free, lively discussions with researchers in various fields about subjects including the current state of research on marine organisms, challenges for and impediments to carrying out research, and new research ideas, as well as to arrange and facilitate distinctive research with the purpose of shedding light on marine ecosystems.

Chapter 44

Lipid Peroxides and Disease

Yasuhito Nakagawa

Introduction

Recent trends in Japan toward a more "Westernized" dietary lifestyle have been accompanied by health problems due to excessive fat consumption. In the 10-year period between 1960 and 1970, human consumption of animal fats quadrupled in Japan, an increase of unprecedented suddenness anywhere in the world. Because excessive fat intake leads to conditions such as myocardial infarction and obesity, the MHLW has cautioned against excessive consumption of animal and vegetable fats in "The Dietary Reference Intakes for Japanese". In addition to cautioning against excessive fat consumption, the MHLW also recommends well-balanced fatty acid intake. Fatty acids can be classified as saturated, monounsaturated, and n-3 and n-6 polyunsaturated fatty acids, and the saturated fats, which make up a high proportion of most animal fats, have been associated with elevated blood cholesterol levels. In this regard, studies have shown that the incidence of myocardial infarction is much lower among Eskimos whose dietary fat comes primarily from fish oil having a high content of n-3 polyunsaturated fatty acids, than among Danes whose diet is focused on meat and other products containing a high proportion of n-6 polyunsaturated fatty acids and saturated fatty acids. These findings suggest that the n-3 fatty acids such as eicosapentaenoic acid (EPA) and docosahexaenoic acid (DHA) play an effective role in preventing myocardial infarction, which has led to a call for greater fish consumption in Japan.

The n-3 fatty acids contained in fish oil are also known to inhibit cancer cell proliferation. The mechanism by which this occurs remains unknown, but recent experiments in nude mice raised on fish oil have shown that n-3 fatty acids promoted apoptosis and inhibited cancer cell proliferation. Detailed research into the mechanism of apoptosis is currently underway across a variety of fields, and numerous inducing and inhibiting factors have been discovered. However, the mechanism of apoptosis induction remains poorly understood, particularly regarding the mechanism of release of pro-apoptotic factors from the mitochondria. The mitochondria are the largest cell organelles producing reactive oxygen species (ROS), and some reports indicate a relationship between ROS and apoptosis induction. Such findings suggest that clarification of the relationship between EPA, active oxygen, apoptosis induction, mitochondria, and inhibition of cancer cell proliferation could be a research project of considerable interest.

Mechanism for EPA−induced apoptosis

We investigated the mechanism of EPA−induced apoptosis, Rat basophilic leukemia cells (RBL−2H3 cells) treated with EPA showed induction of apoptosis that was both concentration−dependent and time−dependent. Apoptosis was similarly induced by the n−3 fatty acid DHA, but not by the n−6 fatty acid linoleic acid or by saturated or monounsaturated fatty acids. Factors such as cytochrome c and apoptosis−inducing factor (AIF), which determine the activation of apoptosis, are released by the mitochondria. Through investigations using agents to inhibit those activating factors, it has become clear that AIF contributes to EPA−induced apoptosis. EPA produced notable increases in intracellular calcium concentration, while apoptosis induction was inhibited by ruthenium red, which blocks mitochondrial calcium influx, suggesting that the flow of calcium into the mitochondria is essential for apoptosis induction. This mitochondrial calcium influx was associated with a pronounced increase in the formation of the active oxygen species hydroperoxide, and apoptosis was blocked by the addition of Mn−TBAP, which eliminates the mitochondrial superoxide. Furthermore, overexpression of the phospholipid hydroperoxide glutathione peroxidase (PHGPx, a unique antioxidant enzyme that eliminates intracellular phospholipid hydroperoxide) in the mitochondria of RBL−2H3 cells was associated with elevated mitochondrial calcium levels, but with no induction of apoptosis. These findings suggest that the formation of hydroperoxide due to elevated mitochondrial calcium, and also the PHGPx substrate phospholipid (phospholipid hydroperoxide) contribute to the AIF released from the mitochondria. In cells overexpressing mitochondrial PHGPx, there is strong inhibition of apoptosis induced by factors such as ultraviolet irradiation, staurosporine, and deoxyglucose, as well as by EPA.

We attempted to identify phospholipid hydroperoxides related to the mitochondrial release of apoptosis inducing factors such as AIF and cytochrome c. In considering mitochondrial release of apoptosis inducing factors, we felt that we should focus on cardiolipin, a phospholipid found specifically within the mitochondria. A permeability transition pore (PT pore), which is a megachannel in the mitochondrial membrane, must be opened in order for the apoptosis inducing factor to be released from the mitochondria. The mitochondrial protein that controls the opening and closing of this channel is the adenine nucleotide translocator (ANT), also known as the ATP/ADP translocator. Cardiolipin, which is essential for ANT activity, contains a high concentration of polyunsaturated fatty acids in comparison to other phospholipids, and is readily oxidized as compared to other phospholipids. Because cardiolipin hydroperoxide was detected in the mitochondria of EPA−treated cells, we decided to use ANT−reconstituted liposomes to review the effects of cardiolipin hydroperoxide on ANT activity. ANT activity was noted in the presence of liposomes including cardiolipin, but was strongly suppressed by the addition of cardiolipin hydroperoxide to the reconstituted liposomes. Our findings indicated that mitochondrial production of cardiolipin hydroperoxide resulted in deactivation of ANT and opening of the PT pores. In addition, cytochrome c has an affinity for cardiolipin in the inner mitochondrial membrane, and bonds loosely to that inner membrane, so in order for cytochrome c to be released, it must be free from the inner membrane to the intermembrane space. When we investigated the affinity between cytochrome c and cardiolipin, we found that affinity to be stronger than between cytochrome c and other phospholipids. However, affinity to the oxide form (cardiolipin hydroperoxide) was extremely weak. Cardiolipin hydroperoxide also greatly affected the stability of the mitochondrial membrane. When the mitochondria were isolated from the cells, the mitochondrial membrane that had integrated with the cardiolipin hydroperoxide became extremely unstable, and mitochondrial swelling and the release of apoptosis induction factors were readily elicited.

These findings indicate that the EPA-induced elevation of mitochondrial ROS levels resulted in oxidation of the unsaturated-fatty-acid-rich cardiolipins, deactivation of ANT, destabilization of the mitochondrial membrane, and release of apoptosis activating factors from the mitochondria, thus inducing apoptosis.

Lipid peroxidation and male infertility

The intracellular enzyme PHGPx, which is responsible for reductionof phospholipid hydroperoxides, is strongly expressed by sperm cells. Our investigation showed this expression to be localized in the midpiece of sperm where almost all mitochondria are present. Just as in lymphocyte formation, the sperm formation process requires active apoptosis in order to maintain cell quality.

As I order to investigate the involvement of mitochondrial PHGPx, an apotosis-inhibiting factor, in sperm formation, we studied PHGPx expression in the sperm of infertile men who were defected the production and mobility of sperm. We found marked reduction in the expression of PHGPx in the sperm of 7 out of 73 infertile male patients. Those patients showed mitochondrial swelling in the sperm, and morphological abnormalities were apparent. Assessment of patients on the basis of WHO classifications for male infertility indicated that, in the category of male infertility patients with low expression of PHGPx, all 7 patients were classified as having serious male infertility with oligozoospermia (low sperm count) and with most of the live sperm showing asthenozoospermia (reduced sperm motility). Approximately 30% of these male infertility patients showed low levels of PHGPx expression, indicating that decreases in the expression of PHGPx seriously interfered with sperm function.

In order to gain a direct understanding of the contribution of PHGPx to sperm formation and function, we created a PHGPx knockout mouse, but the phenotype was embryonic lethal. We then used the Cre/loxP system to create a testicular PHGPx-specific knockout mouse. Mice created specifically to be deficient in testicular PHGPx showed abnormal mitochondrial morphology in sperm and pronounced reductions in sperm count and fertility, indicative of potent effects on differentiation and function in sperm with reduced expression of PHGPx. We believe that the reduced expression of PHGPx in the sperm mitochondria led to elevated mitochondrial levels of phospholipid hydroperoxide, that oxidative stress can develop in sperm due to a variety of factors, and that this oxidative stress can result in the formation of phospholipid hydroperoxides which may be closely related to male infertility.

Chapter 45

Salmonella and Campylobacter Foodborne Disease : An Farm Level Perspective

Masayuki Nakamura

Introduction

Since BSE (bovine spongiform encephalopathy) first appeared in Japan, there has been increasing concern about food safety, and recently a number of problems have emerged regarding "truth in labeling" for a variety of food products ranging from beef, pork, and chicken to chocolate and manju (steamed filled buns). However, these problems have not caused actual injury to humans, for example through hospital admissions due to illness (the "imported gyoza affair", in which contaminated frozen gyoza from China were sold without being labeled as imports, may turn out to involve criminal action, but that is a separate issue from foodborne diseases).

In sharp contrast, when we consider human dietary health issues it is clear that foodborne diseases can result in actual human fatalities. There is nothing more serious than the death of a human being, and this is the point of view from which I approach food safety.

At present, under the Food Safety Basic Law established in May 2003, the Food Safety Commission has been set up independently from governmental institutions (risk-management institutions) such as the Ministry of agriculture, Forestry and Fishery (MAFF) and the Ministry of Health, Labor and welfare (MHLW), which are responsible for risk management through means such as regulation and guidance. The Food Safety Commission was made independent from these other institutions in order to promote science-based food safety administration, and to provide an objective, neutral, and fair-minded scientific assessment of the effects of food on health.

In this article, I will describe the risk assessment targets currently being pursued by the Food Safety Commission joint committee for the review of microorganisms and viruses: (1) Salmonella contamination of chicken eggs and (2) Campylobacter contamination of poultry meat. I will describe the conditions under which such contamination occurs on farms, and will discuss countermeasures.

Salmonella contamination of chicken eggs

Japan currently imports nearly 1,000,000 chicks for breeding use per year. Laying hens that are the offsprings of these chicks maintained during their egg production cycle, and the eggs laid are used for food. If the imported breeding chicks are contaminated with Salmonella enteritidis (SE), this SE will pass transovarially to the next generation of the imported chicks (laying hens) and subsequently contaminate the eggs laid by those hens. Since 1989, the incidence of SE food poisoning increased sharply in Japan, becoming a nationwide problem. In the 1990s, *Salmonella* was in competition with *Vibrio parahaemolyticus* for the most common cause of foodborne iseases in Japan. However, the MAFF and MHW implemented policies including stronger import quarantine measures and "best if used by" labeling for commercially marketed eggs, and a variety of countermeasures were also implemented by Japanese poultry farmers. These measures brought a reduction in *Salmonella* cases beginning in 2000, and during the last few years the incidence of *Salmonella* foodborne disease has dropped sharply in comparison with the incidence of *Campylobacter* and norovirus, the two other most common causes of foodborne diseases.

1) Status of outbreaks on poultry farms

In the 1990s, approximately 15% of poultry farms tested was positive for SE contamination. By 2001, this number had dropped to 3.5%, and today the number is believed to stand at 1%–2%.

2) Countermeasures at poultry farms

a) Vaccines: SE inactivated oil-emulsion vaccine is considered to be effective in inhibiting the production of contaminated eggs. At present there are five companies providing a total of six products in this field, and our research laboratory has participated in the development of each of those products. Since the incidence of contaminated egg production is approximately 1 in every 3000 eggs, it is difficult to prove the efficacy of these vaccines, which currently are officially indicated for "reducing bacterial colonization within the intestinal tract".

b) CE (competitive exclusion) method: Immediately after chicks hatch, they have a gastrointestinal tract that is bacteria-free. Such chicks will die following the oral administration of just a few Salmonella bacteria.

If during this period of high sensitivity the chicks are treated with anaerobic cultures derived from the cecum content of healthy adult chickens, they will develop normal bacterial flora, and Salmonella that subsequently invades into the gastrointestinal tract will be competitively excluded.

This technique is a form of probiotics, and is used worldwide.

c) Feed additives: A number of feed additives are available. However, our research shows that not all of them are effective. One that has proven effective in reducing SE colonization within the intestinal tract is the natural drug Curcuma zedoaria.

3) Primary measures employed overseas (Private-sector-led)

a) England: Breeders and laying hens that are found to be infected by SE are destroyed. Eggs are stored at 20 ℃, and the "best if used by" date is set at 27 days after laying or 21 days after packing.

b) USA: Eggs are stored refrigerated at 7.2 ℃, and are shipped under refrigeration.

Campylobacter contamination of poultry meat

1) Transovarial infectivity of campylobacter jejuni
1986-1997: Vertical infection ruled out
- Attempts to experimentally induce eggshell penetration were unsuccessful.
- Inoculation into the egg albumin resulted in only a few positive chicks after hatching (12/162)
- Breeding hens and their chicks differed in serotype for the Campylobacter strains that were isolated.
- Breeding hens and their chicks differed in RFLP type for the Campylobacter strains that were isolated.

After 2000: Vertical infection confirmed
- Same clonal strains isolated from breeding hens and their chicks
- Isolation of *Campylobacter* from oviducts of breeding hens and laying hens
- Isolation of *Campylobacter* from residue in incubators
- Detection of *Campylobacter* DNA in the intestinal tract of chick embryos during incubation

2) Contamination status of broilers, and shipping to poultry processing plants

Campylobacter infection generally occurred in the chicken flocks when the chickens are approximately 3-5 weeks of age, after which point drinking water becomes contaminated and infection spreads rapidly throughout the flock. At present, 70%-80% of chicken farms in Japan are believed to be contaminated. During shipping to the poultry processing plant, the shipping cargos also become contaminated, which can considerably accelerate the general level of contamination.

3) Routes by which campylobacter invades into farms

No clear theories have yet been formulated regarding exactly how Campylobacter is introduced into poultry flocks. Rather than conducting in-depth research on bacterial infiltration routes into the facility, many researchers have reported on transmission routes for infection within the poultry flocks. In most cases, bacteriologic test findings are negative for the source of water supply and for feed. Vectors of bacterial transmission within the facility may include employees as well as flies and other insects. Further possibilities include vertical infection, manura, small mammals, and environmental factors. Since the air can be eliminated as a sustained source of infection (transmission source), the primary source of infection (transmission source) at farms is most likely to be fecal contamination of the drinking water.

4) Countermeasures

a) Mucosal competitive exclusion: Experiments with mucosal competitive exclusion proved to be effective when using anaerobic culture from mucosal curettage of the cecum from 6-week-old broilers.

b) Probiotics: Probiotics including poultry-specific *Lactobacillus acidophilus* and Streptococcus faecium inhibit intestinal colonization by *Campylobacter* (27% decrease), and also inhibit Campylobacter shedding (70% decrease).

Some reports indicate that the effectiveness of probiotics differs between *Salmonella* and *Campylobacter*. *Saccharomyces boulardii* (a strain of yeast) has been shown to be effective against *Salmonella*, but not against *Campylobacter*. This is probably because *Campylobacter* shows no mannose-specific binding such as seen with *Salmonella*.

c) Prebiotics: There are reports of the effectiveness of adding 4% sucrose and 0.7% caprylic acid (neutral fatty acid) to the feed.

d) Bacteriocins: Bacteriocin OR-7 isolated from *Lactobacillus salivarius* is reported to demonstrate antibacterial activity against *Campylobacter in vitro* and *in vivo*. At first look, this substance appears extremely effective. However, in the case of the in vivo study used day-old chicks, I suspect that the conditions within the gut of such young chicks are not yet well-suited for explosive *Campylobacter* proliferation. Even if chicks of that age are inoculated with 108 CFU, the bacteria will be unable to colonize the as-yet-undeveloped intestinal crypts, and will simply proliferate within the intestinal lumen. If bacteriocin is administered at this stage, the bacteria will be extinguished, but it is questionable whether this is true antibacterial activity. It is important that antibacterial effects can be observed even close to the shipping time, so we would like to see results for testing performed from 5 weeks old until immediately before shipping.

e) Bacteriophages: The *Campylobacter* colonization-decreasing effect of bacteriophages was investigated using 25-day-old chicks in which *Campylobacter* colonization had already occurred. Chicks were orally inoculated with phage and were then monitored for 5 days. Reductions ranging from 0.5 log to 5 log were noted, with the effectiveness varying depending on the combination of *Campylobacter* strain and the phage types.

f) Vaccines: To date, results have been presented for a number of vaccine types including live vaccine, inactive vaccine, recombinant vaccine, and DNA vaccine. These vaccines have been reported to range from completely ineffective through slightly effective to extremely effective, with the latter represented by preparations such as the recombinant vaccine that was created by incorporating *Campylobacter*-related genetic material into attenuated *Salmonella*.

Chicks that had not yet been given either food or water were inoculated orally with this attenuated *Salmonella* recombinant vaccine (10^8 CFU) 4 hours after hatching. Follow-up immunization was administered 2 weeks later, and after an additional 2 weeks, the chicks were orally challenged with a wild-type *Campylobacter*. Chicks were subsequently monitored for increases/decreases in *Campylobacter*. Results showed that the live bacteria count decreased by 6 log or more in the vaccinated group in comparison to the control group. This is due to induction of serum (IgG) and intestinal (IgA) antibody production against *Campylobacter*.

5) Poultry processing plants: Cross-contamination and logistic slaughter

When broilers are contaminated with *Campylobacter* at a farm and that contamination is further exacerbated during shipping, the slaughtering process at the poultry processing plant further adds cross-contamination, which can result in even *Campylobacter*-free broilers becoming contaminated. Generally, the number of bacteria on the poultry carcass is reduced by washing with hot water, increased by the defeathering and evisceration, and then reduced again by chilling. However, cross-contamination can occur in particular in chilling water, with increased contamination both of chicken carcasses and of chicken meat.

In order to avoid such cross-contamination, the Codex Alimentarius Commission created jointly by WHO and FAO to develop food standards has provided a special recommendation that infected poultry should be slaughtered on weekends, or at least at the end of the working day. Countries such as Sweden, Denmark, and Ireland, where the poultry processing plants are relatively small in scale, have adopted and are implementing these recommendations, and are finding them to be quite effective. However, it remains questionable whether such a system is feasible in Japan. Problems include the following. (1) Some producers have schedules in place at least 6 months in advance for chick delivery dates, growing periods, hire dates for chicken catchers, shipping dates, and disinfection of empty chicken sheds. It would not be a major problem if the chickens are slaughtered in the afternoon of the scheduled day, but it is unfeasible in such cases to postpone the slaughter date for several days until the weekend. (2) Large poultry companies may have poultry processing plants at two or three separate locations, so in theory it would be possible to classify incoming chickens as "clean" or "carrier", and to process those groups separately. However, this could lead to considerable confusion if the plants are located at some distance from each other. Furthermore, in order for the processing plant to obtain clear test results and to take appropriate measures, test specimens must be collected 2 weeks before shipping. This means that the processing date is determined 4 to 5 days before shipping (less if PCR is used). (3) In a situation such as we have currently, where the majority of poultry is contaminated, this method will not be very effective. It should, however, prove more useful when we are able to reduce the amount of contaminated poultry (to a contamination rate of 20%-30%).

Conclusion

When *Salmonella* contamination of poultry eggs developed into a national problem, producers became the recipients of societal sanctions. Those sanctions, together with new regulations from the MAFF and the MHLW and greater awareness by producers, have led to continued reductions in the incidence of SE foodborne diseases.

In contrast, *Campylobacter* contamination of poultry meat is completely unregulated. The issue has received almost no attention from the mass media, and producer awareness is low. Currently, risk assessment is being conducted by the Food Safety Commission. As a member of that Commission, I am monitoring the Commission's progress and stating my opinions.

The primary issues regarding *Campylobacter* foodborne diseases are (1) contamination at the poultry farm, (2) cross-contamination at the poultry processing plant, and (3) inadequate cooking of poultry and the consumption of raw meat such as chicken gizzard, liver and so on. In this article, I have not addressed the issues at the consumer level. However, although the foundation of foodborne disease countermeasures lies in the progression of Farm→Poultry Processing Plant→Meat Processing and Transport→Consumption, *Campylobacter* does not proliferate at the meat processing and transport stage because of the microaerobic environment there, so assessment is easily implemented. However, at poultry farms it is difficult to specify the source of infection and of transmission, which makes it much more difficult to prevent bacterial invasion. There is also the additional difficulty, not commonly seen overseas, of the Japanese cultural proclivity for eating raw gizzard and liver and other forms of raw meat. More time will be required to resolve these problems, and that resolution will require the cooperation of everyone involved with the process, from the farm to the poultry processing plant, the meat processing and transport industry and the consumer.

Chapter 46

Problems Involving Arsenic : Induced Health Effects in People Who Consume Much Seaweed

Hiroshi Yamauchi

Arsenic and health damage in recent years

 1. Arsenic, representative of which is arsenic trioxide, is a typical poison that has been used for homicide and suicide from Middle Ages Europe until modern times. Arsenic trioxide is a byproduct of copper refining, and is also made from arsenic sulfide. From 50,000 to 60,000 t are used annually in Japan, with primary applications being LCD glass substrates and semiconductors with arsenic compounds. There are concerns about health effects from occupational exposure. Because in the past much arsenic was used in applications such as pesticides, herbicides/defoliants, and wood preservatives (chromium− copper−arsenate, or CCA), there is much soil seriously contaminated with arsenic, and much discarded CCA−treated wood. There were very many cases of skin and lung cancer owing to occupational arsenic exposure and owing to the inorganic and organic arsenic compounds used in pharmaceuticals (such as tonics and compounds used to treat syphilis, infections, and parasites). Currently the spotlight is trained on the apoptosis effect arsenic trioxide has on cancer cells, and it is now used in drugs for treating acute leukemia (acute promyelocytic leukemia, APL).

 2. About 70 years ago, Japan manufactured 7,000 t of chemical weapons, many of which contained arsenic. Even now, 4,000 t (300,000−400,000 artillery shells) remain in China, and 3,000 t in Japan. Currently in Japan, about 160 people poisoned by arsenic chemical weapons are receiving treatment, and in China, there are accidental deaths and injuries related to chemical weapons.

 3. In recent years, mainly in Asian countries, pumped wells (15−20 m) have been dug for the purpose of preventing water−borne communicable diseases and securing water for agriculture. However, well water is contaminated with naturally high concentrations of inorganic arsenic compounds. According to statistics of international agencies, there are about 80 million victims of chronic arsenic poisoning, including potential victims, and there is a

worsening trend in East Asia.

4. In July 2004, the UK Food Standards Agency (FSA) recommended a ban on the consumption of hijiki seaweed in the UK out of concern for the health effects of the high concentrations of inorganic arsenic in hijiki, and UK actions in this regard were widely reflected in other Western countries. As a result, hijiki is now eaten only in Japan. In March 2004, the author and others prepared a report for the Food Safety Commission entitled "Report on a Review of Basic Documents on the Assessment of Arsenic in Hijiki". In the interests of caution, basic research has also been started on the health impacts of arsenic compounds in other seaweeds (such as kombu, wakame, and nori).

Arsenic has a mechanism of action absent from other metals and metaloid elements. Arsenic's chemical structure and chemical morphology strongly influence the biological effects of arsenic compounds on humans and these effects are not determined solely by "amount of exposure" or "amount ingested". Because arsenic compounds are intimately connected to our food and living environments, it is essential to manage the risk of acute and chronic poisoning and carcinogenicity. A problem related to this, and which is now the focus of attention, is the health effects of arsenic compounds from excessive consumption of seaweeds.

Chemical forms and concentrations of arsenic in seaweeds

An environmental characteristic of arsenic compounds is that many are found in higher concentrations in marine organisms than in terrestrial flora and fauna. While the concentrations of arsenic are low in plants cultivated and produced in the soil and in livestock products, the concentrations of total arsenic detected in some seaweeds, fish, and shellfish range from several to several hundred ppm (μ g/g), and the difference between food groups from terrestrial and marine organisms is perhaps 1,000 times. Even among marine organisms, there are large differences between the chemical structures of arsenic compounds found in seaweeds and those found in fish/shellfish. The latter contain the trimethyl arsenic compound arsenobetaine (AsB), which has no acute toxicity, and an LD_{50} value of 10μ g/kg, which is 1/300 that of arsenic trioxide (LD_{50}: 0.03 g/kg). AsB has low tissue affinity, it is quickly eliminated along with urine, and its half-life is 3-5 h. By contrast, the concentrations and chemical structures of arsenic in seaweeds are complex. The arsenic detected in seaweeds can be roughly divided into the two categories of inorganic arsenic and dimethyl arsenic compounds. Inorganic pentavalent arsenic is detected in the brown alga hijiki, and the concentration of inorganic pentavalent arsenic in commercially sold dried hijiki is high at 10-100 μ g/g. The arsenic detected in other seaweeds (such as kombu, wakame, and nori) comprises mainly arsenosugars (As-Sug), which include many forms . At least 20 kinds have been confirmed, and they are present in high concentrations of several ppm to several tens of ppm. Limited research performed on mammals shows that As-Sug are metabolized and decomposed by the liver and intestinal bacteria, changed into dimethylarsinic acid (DMA), and eliminated in the urine. This DMA is the same as the metabolite of arsenic trioxide. Although the LD_{50} values of As-Sug have not been precisely calculated, because that of the metabolite

DMA ranges between 1,300 and 1,500 mg/kg, there are likely no concerns about acute toxicity. However, because arsenosugars are transformed into DMA in the body, the problem is not that of impacts from acute and chronic toxicity, but of carcinogenicity risk.

Field study

"Arsenic ingestion and health effects in people who consume much seaweed"

Research study and method: Our study area was the Iwate Prefecture coastal zone, and subjects totaled 119 people (59 men 51.4 ± 14.0 years and 60 women 52.1 ± 12.9 years), comprising 79 people who grow wakame and engage in abalone and sea urchin aquaculture, and their families, and 40 people who engage in sea squirt aquaculture, and their families. We conducted a meal study with the "duplicate portion system, recovered samples of all the food and beverages for subjects separately for breakfast, lunch, dinner, and snacks, and took spot urine samples on the same days. At the same time, we gave questionnaires to subjects on their dietary lifestyle and health condition, and conducted health examinations. We treated meal samples with the nitric acid ashing method and obtained the total arsenic concentration with inductively coupled plasma atomic emission spectrometry (ICP-MS). We measured arsenic compounds in the blood according to chemical species (inorganic arsenic (iAs), monomethylated arsenic (MMA), dimethylated arsenic (DMA), and AsB by pretreating samples with alkaline thermolysis and then using the cold trap-reduction-atomic absorption method. To determine oxidative DNA damage, we measured the concentration of 8-hydroxy-2'-deoxyguanosine (8-OHdG) in the urine with the ELISA method. Arsenic and 8-OHdG in urine were corrected with the concentration of creatinine in urine. As our control group, we used the concentrations of arsenic and 8-OHdG in the urine of 248 adult men and women living in six areas of Japan, and who were found healthy by medical examinations.

Daily amounts of seaweed, fish, and shellfish from meals

In this study, none of the subjects ate hijiki, while most ate wakame. While in general a Japanese eats about 14 g of seaweed daily, the average for the 119 subjects in the experimental group tended to be higher at 21.8 ± 26.6 g Further, there was a pronounced difference between men and women, with women eating the large amount of 28.2 ± 28.2 g, and men 15.2 ± 23.3 g. The experimental group's amount of fish and shellfish was 108 ± 71.7 g, which was not different from the national average (about 106 g). However, we noted that in the experimental group the men (121 ± 75.5 g) tended to eat more than women (96.6 ± 66.2 g).

Daily amount of total arsenic ingested from food

In Japan's culinary culture the Japanese eat more seaweed, fish, and shellfish than other ethnic groups, and from this it has been known that they tend to inevitably ingest more total arsenic from food. In 1992, the author used the same "duplicate portion system" to conduct a survey of city dwellers (35 men and women in Kawasaki City) and found that their total arsenic consumption was $195 \pm 235 \mu g/day$. The result obtained by Mohri et al. in Fukuoka City was similar at $202 \pm 143 \mu g/day$. In this study, the average daily total arsenic consumption of the 119 subjects was $336 \pm 417 \mu g/day$, which was nearly twice the amount for city dwellers. The average value for women, 374 ± 546 μg/day, tended to be larger than that for men, 297 ± 216 μg/day, but no statistically significant difference was found between them.

Urine arsenic concentrations and chemical forms

The average urine arsenic concentration by chemical species of the 248 healthy Japanese was 149 ± 129 μg/g creatinine (iAs, 2.4%; MMA, 1.3%; DMA, 26.8%; and AsB, 69.1%).

This study found that the average urine arsenic concentration of the 119 subjects was high at 304 ± 391 μg/g creatinine (iAs, 0.5%; MMA, 2.5%; DMA, 36.8%; and AsB, 60.2%), which was about twice the average value for the 248 people in the control group. The average value for men was 251 ± 202 μg/g creatinine (iAs, 0.6%; MMA, 3.4%; DMA, 36.7%; and AsB, 59.0%), and that for women was 357 ± 510 μg/g creatinine (iAs, 0.5%; MMA, 1.9%; DMA, 36.7%; and AsB, 61.0%), revealing that the average value for women is higher than that for men.

During the study period women ate much wakame, whose primary type of arsenic is As-Sug, and the fact that As-Sug's metabolite is DMA could explain the rise in urine DMA concentration. Hence, the results showed a significant correlation ($r=0.357$) between the amount of seaweed eaten and urine DMA concentration. There was also a significant correlation ($r=0.375$) between the amount of fish and shellfish eaten and urine AsB concentration.

Arsenic ingestion and oxidative DNA damage

Currently 8-OHdG is widely recognized as an effective marker of oxidative DNA damage. The average urine concentration of 8-OHdG in the 248 healthy people was 15.4 ± 5.60 ng/mg creatinine. The value for the 128 men was 15.2 ± 5.19 ng/mg creatinine, while that for the 120 women was 15.6 ± 5.49 ng/mg creatinine, revealing no discernible differences for gender or age (ages 20-65).

The average urine 8-OHdG concentration of the 119 study subjects was 17.3 ± 6.79 ng/mg creatinine, which tended to be somewhat higher than that of the control group, but we found no statistically significant difference between the groups ($p<0.01$). Values for men and women were, respectively, 17.6 ± 6.98 ng/mg creatinine and 17.0 ± 6.64 ng/mg creatinine (no significant difference). We found no statistically significant difference between the urine 8-OHdG concentration of the 119 study subjects and the amounts of seaweed, fish, and shellfish consumed, and there was no correlation between the urine arsenic concentration by chemical morphology and urine 8-OHdG concentration.

Conclusion

It is widely acknowledged in society that seaweed is a beneficial food from a dietetics perspective, but as this research showed, the daily total arsenic amount ingested by fishery workers and their families, who eat much seaweed, fish, and shellfish, tends to be higher than that of city dwellers. There is particular concern with regard to the significant correlation between the amount of seaweed eaten and urine DMA concentration. This DMA is the same as the main metabolite of inorganic arsenic, and its general toxicity and carcinogenicity make it an important subject for future research. On the other hand, excessive ingestion of AsB from eating fish and shellfish is not considered to present any problems in the way of biological effects because this arsenic is nontoxic.

I feel a need to take good care of Japan's culinary culture, in which hijiki and other seaweeds are eaten, but in consideration of arsenic's toxicity, I think it is necessary mainly in the case of pregnant women, infants, and small children to look elsewhere for a source of minerals instead of using seaweeds as we have been. At the same time, even for adults in general to eat large amounts of seaweed at one time means retaining excessive toxic arsenic in the body, which could increase the risk of cancer. People should be careful not to eat too much at one time.

The Role of the Center for Genetic Studies of Integrated Biological Functions in Agromedicine: The Kind of Laboratory Animal Facility Needed for Future Medical Research

Nobutaka Shinohara

It goes without saying that basic medical research has in large measure been supported by animal research, but progress in reproductive engineering techniques and the augmentation of basic data on genomes have exponentially boosted the importance of the role that will be played by animal experiments in medical research henceforth. Owing to this trend, laboratory animal facilities not only find themselves required to play a bigger role, but are also pressed to make qualitative changes. From the late 20th century and into the start of this century, the entire genome base sequences of animal species including humans and mice have been determined. This is like having completed a dictionary, and our era (the post-genome era) is now one in which scientists can proceed with biological research with a dictionary in hand. No matter how well one understands the meanings of individual words with the dictionary, that is merely the basis for understanding a foreign language. Similarly, life is an extremely complex phenomenon that can never be understood by just combining simple theories of materialism. In biomedical research, gene functions must be understood in the context of high-order biological control programs such as development, differentiation, and homeostasis in multicellular organisms. Laboratory animals in which certain genes have been transferred, destroyed, modified, or otherwise changed are of great use in analyzing the physiological functions, roles, and control of expression of individual genes (or gene products), as well as the morbid physiology caused by their malfunctioning. Laboratory animals are indispensable to medical research, and they will likely continue holding a permanent place in future biomedical research. In view of this situation, the role and function expected of laboratory animal facilities is now far more extensive than before.

Nearly all the genetically modified animals that are created one after the other are supplied by sources other than designated businesses, but from now on a required condition for laboratory animal facilities will be to provide an environment facilitating their smooth introduction and use. However, at many of the animal facilities now being created, in order to maintain cleanliness there are extreme limits on the exchange of animals with other facilities, but this is harmful because it imposes severe limitations on research. University laboratory animal facilities are not pretty museums for animals, but rather facilities that support research. Improving cleanliness is seen as being incompatible with facility convenience, but need this be really so?

Efforts made at the Kitasato university school of medicine's center for genetic studies of integrated biological functions

The Center for Genetic Studies of Integrated Biological Functions, which is a facility of the Kitasato University School of Medicine, is a laboratory animal facility whose design and operation are much different from the usual thinking in order to address this change in demand. Its main characteristics are as follows.

(1) It not only maintains a high level of animal cleanliness, but also has a "distribution area" to facilitate smooth distribution. This area can directly accept animals that are somewhat problematic with regard to cleanliness and use them in experiments. When it is necessary to maintain at least three generations, the area has the animals submitted, cleans them, puts them in the SPF (specific pathogen–free) area, and at the same time preserves frozen embryos.

(2) Reproductive engineering techniques are integrated into everyday operations, and animal cleaning, embryo freezing, and other activities have become a part of general services. This has facilitated the use of various genetically modified animals that are created in other facilities.

(3) Using our distribution area with these functions as a bulwark, we made operating rules for the SPF area that allow no exceptions whatsoever. The SPF area never accepts animals other than those from designated businesses, animals born in the SPF area, or animals cleaned in this facility.

(4) In full consideration of user needs, we achieved a quantum leap in convenience by designing traffic lines with very careful thought and rejecting any rules that might be considered unnecessary superstitions.

(5) Users are registered under each year's research plan, and have card keys that open only those areas for which they are registered.

The facility has been operating in this manner for six years, and everything is proceeding very smoothly and according to expectations. In the initial period when operating rules were not thoroughly implemented there were several instances in which pathogenic microorganisms made their way into the SPF area, but because all instances were caused by bringing in contaminated animals, thorough implementation of operating rules completely eliminated problems. This is a compact and highly convenient facility, and there are currently few other facilities like it in the world because of the high quality of its animal care environment, including the cleanliness of its SPF area. We believe it can serve as a model for the future design and operation of the animal facilities of universities and other research institutions.

Chapter 48

Possibilities of Probiotics as a Bridge for Agriculture–Medicine Collaboration

Takao Mukai

Introduction

Underlying the consumer health boom in Japan during recent years is the sale of many probiotic products including fermented milk products. Many of those products have obtained permits to label themselves as "foods for specified health uses". Probiotic products are also increasingly being used as additives for livestock feed. At the same time, the various effects that probiotic products have on biological functions and pathogenic microorganisms are being revealed by science, and even in the domain of medical science, attention is focusing on probiotics as a supplementary treatment.

Behind the spotlight on probiotics is the store of scientific evidence that the makeup of the intestinal microbiota is heavily involved in the host's health, and from the perspective of preventive medicine, work has been done on the development of probiotics that improve intestinal microbiota balance. Further, the latest research is now finding that probiotics not only have just intestinal microbiota-regulating effects, but also that probiotics themselves have functions which directly improve host health. From this perspective, it is perhaps no exaggeration to say that probiotics are a bridge that links "food and medicine", in other words, "agricultural science and medical science". At this symposium I will describe the basic concept of probiotics and the related prebiotics, synbiotics, and biogenics. I will also give an overview of the effectiveness of probiotics and other preparations used in not only foods but also as additives in animal feed, and explore the possibilities of probiotics as a subject for research and education in agromedicine.

The Role of intestinal microbiota

It is crucial to understand the relationship between the intestinal microbiota and health when exploring probiotics and prebiotics. It is thought that several hundred species of bacteria live in the human intestine, and that 1 g of stool contains at least 1 trillion bacteria, which make up the so-called intestinal microbiota. The intestine's indigenous bacteria maintain a certain flora while living in symbiosis and competing with one another, and they are part of an important biological defense system that functions to keep out exotic pathogenic bacteria. The intestinal microbiota also plays in important role in the development and maintenance of the intestinal immunity system, and it is involved in the onset or suppression of intestinal disorders including infections, allergies, and inflammation. Recently the spotlight has been trained on the role of the short-chain fatty acids that arise in the fermentation process in the colon. Of the short-chain fatty acids it is mainly butyric acid that is consumed in the colon, and it has been shown to have various physiological functions such as facilitating the normal proliferation of colonic epithelial cells, facilitating intestinal movement, governing inflammatory responses, and promoting the absorption of electrolytes. At the same time, butyric acid induces the apoptosis of cancer cells.

Once the intestinal microbiota is established, it is comparatively stable, but if the microbiota is disturbed for some reason, it is possible that an endogenous pathogenicity will be manifested. Further, some species of intestinal bacteria produce carcinogens, bacterial toxins, and putrefied substances such as ammonia and hydrogen sulfide, so if such bacteria continue to be dominant in the intestinal microbiota, it is possible they will have adverse effects such as damaging the intestines or causing colon cancer or other colon disorders. Scientists have shown how the intestinal microbiota is intimately linked to the health of humans (hosts), and attention has focused on probiotics from the perspective of controlling intestinal flora to maintain normalcy.

Probiotics, prebiotics, synbiotics, and biogenics

Probiotics is a concept that contrasts with antibiotics. Its root is an ecological term that signifies probiosis among organisms, and the definition proposed by Fuller is widely accepted: "A live microbial feed supplement which beneficially affects the host animal by improving its intestinal microbial balance"[1]. Subsequent research has found bacterial strains that have various beneficial effects even if the bacteria are dead, and now these are understood to be "the constituents of microbial supplements or microorganism cells which exhibit effects that are beneficial to the host's health and to maintaining and improving it". The concept of prebiotics is important from the perspective of "improving the intestinal microbiota balance". Prebiotics are defined as "food ingredients that stimulate or activate the growth of certain useful bacteria living in the intestine". Oligosaccharides, yeast extract, and other substances are used. Using prebiotics and probiotics together to improve the intestinal microbiota is synbiotics, another concept that has become widely known. Additionally there is biogenics, proposed as a new concept by Mitsuoka[2], who offers this definition: "Food components derived through microbial activity which provide health benefits without involving intestinal microbiota".

Lactic acid bacteria and bifidobacteria gaining prominence as probiotics

The bacterial species currently being used or researched as probiotics include those from among these genera: *Lactobacillus, Enterococcus, Bifidobacterium, Lactococcus, Clostri-*

dium, and *Bacillus*. Of these, lactic acid bacteria and bifidobacteria are getting the most attention and research because not only are these species found among the intestinal microbiota of healthy people, they have long been consumed in food throughout the long history of humanity, and are therefore thought to be very safe. Many lactic acid bacteria species are also found on the US Food and Drug Administration GRAS (Generally Recognized as Safe) list.

Use of probiotics and prebiotics by humans, and problems

Permission is granted to label a product as a food for specified health uses only when it promises certain healthful effects from the perspective of medical science and dietetics. When a probiotic has been approved as a food for specified health uses, it can be labeled as a "food that regulates one's abdominal condition" (intestinal regulation effect), and this is of great significance. This is because one can expect probiotics to manifest various healthful functions such as by proliferating in the intestines, suppressing the proliferation of putrefying bacteria in the intestines, regulating immune functions, and inactivating cancer-producing enzymes.

Recently, many healthful effects have been reported on the strength of experiments using *in vitro* cells and laboratory animals and randomized human ingestion tests with placebo controls based on the assessment methods used for drugs. Such healthful effects include infectious disease prevention, allergy prevention, mitigation of inflammatory bowel diseases, and anti-cholesterol effects. Meanwhile, a problem that must be solved is that probiotics have different effects depending on the differences in the intestinal microbiota between individuals. It is difficult to expect that one kind of probiotic will have the same effect in all people. There are now hopes for the development of "tailor-made (order-made) probiotics" that take advantage of the useful intestinal lactic acid bacteria (bifidobacteria) found in all people, and for the development of probiotics adapted to individuals, achieved by advances in research in areas such as analyzing the meta-genomes of an individual's intestinal microbiota. Further, for certain disorders there is promise that probiotics will be effective as supplementary treatment methods in a clinical context. In the United States in particular, there is a trend toward recommending supplementary treatment with probiotics because they are considered effective for treating infectious diarrhea, preventing antibiotic-associated diarrhea, and treating atopic eczema in infants and small children. My personal view is that, in consideration of the characteristics of probiotics, it is difficult to expect efficacy which exceeds that of a supplementary treatment. I think it is preferable to use probiotics from a preventive medicine point of view.

The idea behind prebiotics is to promote the proliferation of useful bacteria among those in an individual's intestinal microbiota. Indigestible oligosaccharides are used as a primary material because they are hard for human digestive tract enzymes to break down, and therefore make their way to the colon and are used by bacteria.

Their function is reported to have the same effect as probiotics. The thinking underlying prebiotics originally arose from the fact that bifidobacteria are more dominant in the intestinal microbiota of breast-fed infants than in infants fed baby formula. Very recently it was discovered that the real bifidobacteria proliferation factor is a disaccharide with the structure of lacto-N-biose, which is found in breast milk. There are expectations for future application in foods, including clinical use for infants who need formula.

Making even more advantageous use of the functions of probiotics and prebiotics is possible through mutual cooperation by experts in the agricultural science field, who are food professionals, and experts in the medical science field, who can prove efficacy in humans from a medical perspective.

The use of probiotics for livestock and poultry

To promote growth and improve feed efficiency, currently probiotics is approved for use as feed additives. In livestock and poultry production, probiotics are closely watched as substitutes for antibacterial substances (antibiotics and artificial antibacterial preparations), and their use is growing. One underlying factor is concern about the emergence of drug-resistant bacteria due to the use of antibacterial feed additives. Antibacterial substances have been used around the world not only as veterinary drugs for treating disease in livestock and other animals, but also as growth promoters. Especially when used for the latter purpose, there have been concerns that drug-resistant bacteria in animals would be selected for by long-term use, and for such reasons the EU totally banned the use of antibacterial substances in animals for growth promotion starting in January 2006. At the same time, there is dissent in the livestock industry to the idea that contamination by drug-resistant bacteria will widen; for that reason there should perhaps be science-based quantitative risk analyses of the connection between the use of antibacterial feed additives in animals and the emergence of drug-resistant bacteria. In any case, except for a few developed countries, there is a turn toward banning or reducing the use of antibacterial substances for purposes other than disease treatment, making it urgent to develop substitutes for feed additives containing antibacterial substances.

The efficacy of antibacterial substances for promoting livestock growth appear to be mainly mechanisms that work through the intestinal microbiota, such as by (1) preventing potential infections in juvenile animals, (2) improving intestinal metabolism, and (3) suppressing harmful bacteria. However, in probiotics and prebiotics researchers have found effects for growth promotion and feed efficiency improvement that work by the same mechanisms. There are also suggestions that the use of probiotics in livestock and poultry is effective not only from the perspective of reducing the amounts of antibacterial feed supplements used but also from the perspective of preventing illness in animals, eliminating pathogenic microorganisms that create public health problems, and preventing offensive odors by suppressing production of putrefaction products in the intestines. These applications too are receiving attention.

At the Kitasato University School of Veterinary Medicine, we are promoting a materials recycling system of animal husbandry, and we believe that by introducing probiotics into the cycle, we will make an even greater contribution to maintaining and improving animal health, mitigating environmental problems, and promoting food safety. Keeping animal intestines healthy will help bring good-tasting livestock products to consumers, and indirectly help maintain our health.

Train the spotlight on the infection – prevention effect of probiotics and fermented milk products !

The author's laboratory is focusing on understanding the role of lactic acid bacteria in the intestines and on the infection-prevention effect of probiotic lactic acid bacteria. We are conducting molecular-level research in several areas from the standpoint of obtaining scientific substantiation.

One research area is anti-adhesion therapy that focuses on the competition that arises between lactic acid bacteria and pathogenic microorganisms at adhesion sites. So far we have thrown light on a number of adhesion-inhibiting mechanisms including those of *Lactobacillus kitasatonis*, which suppresses the cellular attachment of *Salmonella*; *Lactobacillus reuteri*, which inhibits the adhesion to receptors of *Helicobacter pylori*, a bacterium suggested to cause stomach cancer; and *Lactobacillus gasseri*, which arrests the adhesion to cells of the food poisoning bacteria *Campylobacter jejuni*.

In addition, because of the problem of drug-resistant bacteria, we are also looking for new antimicrobial agents produced by lactic acid bacteria.

Conclusion : probiotics and agromedicine

Lactic acid bacteria and bifidobacteria play the leading probiotic roles, and it has been shown that the probiotic effects of these bacteria are dependent on the bacterial strain. This makes it necessary, for example, to develop efficient screening methods to find strains with useful functions. Nevertheless, such efforts and the development of probiotics for human use will make no progress in illuminating probiotic functions without the cooperation of researchers in the field of medical science. On the research level, agriculture?medicine collaboration has already achieved considerable progress, and that collaboration will probably grow stronger. The Kitasato University School of Medicine and the School of Veterinary Medicine's Department of Animal Science are considering providing agromedical education starting the next academic year, which would be the first such attempt in Japan. They are now discussing content, and they plan to offer some items with probiotics as the subject matter because probiotics are thought to be the best subject to link food/environment with medicine.

1) Fuller R., Probiotics in man and animals. J. Appl. Bacteriol. 66, 365-378(1989).
2) Mitsuoka T., Significance of dietary modification of intestinal flora and intestinal environment. Bioscience Microflora 19, 15-25(2000).

Chapter 49

Possibilities and Limitations of Functional Foods

Keizo Arihara

Introduction

If you say "foods that are good for your health", I think many people will think of "health foods". But there are considerable doubts about how good for you these health foods really are. In truth, there have been not a few stories about people who suffered health damage because of health foods. A study by the Japanese Ministry of Health, Labour and Welfare (MHLW) released August 2008 found that pharmaceutical ingredients whose unauthorized sale is banned by the Drugs, Cosmetics and Medical Instruments Act were detected in 15% of health foods that made claims of physical invigoration. A 2005 study by the Tokyo Metropolitan Government found that 85% of the health foods on the market either violated or were suspected of violating laws covering their labeling and advertising (Drugs, Cosmetics and Medical Instruments Act; Law to Prevent Excessive Premiums and Unreasonable Representations; JAS Law; and Health Promotion Law).

At the same time, many people have certain expectations for the efficacy of supplements and other health foods. In a recent private study (subjects were 500 men and women aged 50 – 79), 48.8% said they "ordinarily take supplements", and of those, 87.3% said they "take them every day". Probably there are also many people who eat health foods while feeling some doubts about them. Here I would like to offer some ideas for discussing health foods and functional foods, asking what capabilities we can expect of them and to what extent (possibilities and limitations).

Are health foods necessary?

When asked at a forum "Are health foods really necessary?", Dr. Keizo Umegaki of the National Institute of Health and Nutrition replied, "I don't think most people need them".

Dr. Akira Murakami at Kyoto University says, "Many people use health foods because they're lazy, and many people who depend on health foods are cheap."

I too have doubts about using health foods that are too convenient, and I think that in the first place it is above all else important to consider the issue of "food and health" in terms of one's overall dietary lifestyle. Expressed simply with that in mind, my idea is "health foods are unnecessary, but functional foods are necessary". We often hear the term "functional foods", and I will explain this important term in the next section.

Health foods and functional foods

I am a functional foods researcher, and it grieves me to have functional foods and health foods put in the same pigeonhole. Unfortunately, however, many people think of them as more or less the same thing. The following diagram shows how I see in my mind the relationships of ordinary foods, health foods, functional foods, foods for specified health uses, and pharmaceuticals.

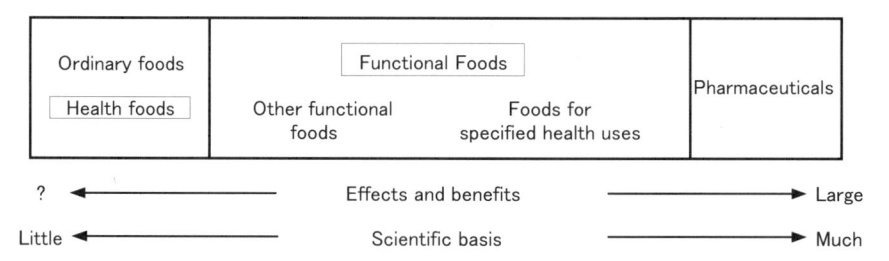

```
┌─────────────────┬────────────────────────────────────────┬────────────────┐
│ Ordinary foods  │         Functional Foods               │ Pharmaceuticals│
│                 │                                        │                │
│ Health foods    │  Other functional     Foods for        │                │
│                 │      foods         specified health uses│               │
└─────────────────┴────────────────────────────────────────┴────────────────┘

   ?        ◄─────────── Effects and benefits ───────────► Large
   Little   ◄─────────── Scientific basis    ───────────► Much
```

But when adding legal circumstances, no clear distinction can be made between health foods and functional foods. I want readers to see the difference between health foods and functional foods as the presence or absence of scientific basis. While the government view of health foods is "broadly, foods in general that are sold and used as foods which contribute to the maintenance and improvement of health", this has no scientific basis. As things stand now many health foods are "foods which merely give the impression of health". Herein we find the reason that I want to make a distinction between health foods and functional foods. I should add that foods for specified health uses may be considered functional foods that have met a number of conditions and have legal approval (as of August 31, 2008, 797 items had permission to be labeled as such).

"Food and health" for what purpose?

The health food market is currently seen to be about ¥1.6 trillion. Supplements alone account for about ¥700 billion. The over-the-counter drug market (cold remedies, digestive medicines, eyedrops, tonics, vitamins, and others) is about ¥610 billion, which is arguably quite large. About 10 years ago the OTC drug market was approaching ¥1 trillion, but in 2005 it yielded first place to foods for specified health uses.

This was big news to the parties concerned. One cannot deny that market appeal (size and growth promise) was one reason for the appearance of many dubious health foods, but recent household finance surveys indicate that consumer health food expenditures have peaked, and there are whispers that the market has expanded to its limit. I hope this will be for the best by leading to the elimination of inferior health foods and the development of excellent functional foods.

The media too have all covered "food and health". Of course, this is because of viewers' and readers' needs, but it seems that especially television has created some programs merely as a way of bolstering ratings. Even if viewers might see them as "informational programs" and use the information with gratitude, if producers make them as "variety programs" without serious thought, the result is tragic. TV programs are very influential, and I strongly sense the presence of TV such as when taking questions after lectures to the general public. Although considerable caution has been exercised since the "white kidney bean incident" (where almost 160 people were struck with vomiting and diarrhea after following a weight-loss plan involving white kidney beans that was broadcast by a major TV network in Japan), there are concerns about what people do once they forget the pain.

Although I have strayed a bit from my subject, it appears that current problems involving health foods arose in part because the industry and the media have not always treated the matter of "food and health" with sincerity.

Foods for special dietary uses too are functional foods

The appearance of "hypoallergenic rice" (low-allergen rice) was no doubt very good news for people with rice allergies. This rice was put on sale in 1993 by Shiseido with the product name "Fine Rice"; as the first food for specified health uses, it is a functional food that also has historical significance. Fine Rice is now positioned not as a food for specified health uses, but as a "food for ill people" within the category of "foods for special dietary uses". Some of the categories of foods for special dietary uses established by law are "foods for ill people", "powdered milk for pregnant and nursing women", "infant formula", and "foods for the elderly". As of March 2008, 505 such foods have been authorized.

Foods for the ill and other foods for special dietary uses have a somewhat special position as functional foods, and there is a very great need for them. I myself was involved in the development of foods for those with difficulty chewing and swallowing (foods for the elderly), and I think that the development of such foods has great social significance.

Dietary lifestyle and nutritional education

In 2007, much discussion arose around a book (Boushoku no Jidai, Fusosha Publishing) which recounted the story of an elementary school child who said matter-of-factly, "I chewed gum for breakfast". While a bit hard to believe, it is apparently not unusual to hear about such things on school campuses. This is perhaps not unrelated to the trend in which adults take the easy route toward nutrition by depending on health foods. Although there are pros and cons with the Basic Law on Nutritional Education, which entered into force in 2005, the current situation perhaps requires the government to now consider dietary lifestyles.

At the First Agromedicine Symposium (March 10, 2006) Professor Yoshiharu Aizawa, dean of our School of Medicine, delivered a talk entitled "Agromedicine as Seen from Medical Science". In that talk he emphasized the need for nutritional education and said, "I perceive the need to encompass not only 'food' but the entirety of 'food and eating', and to systematize conventional food science as food and eating science". While there is great significance in functional foods, considering "dietary lifestyle in its entirety" always takes precedence, and that is one of the major goals of agromedicine. At the Second Agromedicine

Symposium (October 13, 2006) Professor Tomiharu Manda of our School of Veterinary Medicine delivered a talk titled "Conservation Livestock Farming ? From Production to the Hospital" in which he described how safe, worry-free beef raised on 100% farm-grown pasture at the Kitasato University Yakumo Farm is used in meals given to patients in Kitasato University Hospital. Professor Manda remarked, "Here we can see the collaboration of agriculture, the environment, and medicine". Efforts underway at Yakumo Farm are communicated to citizens through public lectures and other means, which could also be seen as nutritional education activities conducted by Kitasato University.

Scientific basis and communication to consumers

A scientific basis to establish the healthful effects of functional foods is needed in order to clearly differentiate them from products merely portraying themselves as health foods. At the First Agromedicine Symposium, Professor Takafumi Kasumi of the Nihon University College of Bioresource Sciences emphasized "nutrigenomics and tailor-made foods" in a talk called "Human Health and Functional Foods". Comprehensive analyses using DNA microarrays to examine the effects of food constituents on gene expression are now frequently conducted, and it seems we have made a certain amount of progress in understanding the relationships between food functions and scientific evidence. I am exploring the function analysis of peptides formed by the breakdown of food proteins, and their applications in food. I have incorporated analyses using DNA microarrays, and the results look very promising.

Despite progress in technologies to obtain a scientific basis, one feels the limits of functional foods in the difficulty of communicating the science to consumers. At present, a few functional foods are authorized to have certain effects and benefits listed on their labels because they have obtained permits from the MHLW as foods for specified health uses. However, for functional foods other than those for specified health uses, there is no way to use the data that scientists have gone to the trouble of assembling. Because some foods are hard to approve as foods for specified health uses, it seems we need something more than the "foods for specified health uses" system.

Manufacturers take great pains in naming and labeling their products in consideration of the Drugs, Cosmetics and Medical Instruments Act and other related laws and notifications from the MHLW. Nevertheless, product concepts are not necessarily communicated well to consumers; some products invite misunderstanding, and in some cases give people dubious impressions. I do not agree that the system should be more complicated, but the present system still needs more work. The "foods for specified health uses" system became effective in 1993, and in 2001, the "food with health claims" system was launched. This further brought into existence "foods with nutrient function claims", but very few consumers understand these terms correctly.

Incidentally, one often sees foods labeled "patent pending" or "patented", but in many cases, we do not know what exactly is or will be patented. Additionally, the "revised examination criteria for novelty and inventiveness" issued in recent years by the Japanese Patent Office make it hard for the "invention of use" of functional foods to be recognized, and they are an impediment to developing functional foods.

Some thoughts on pet food development

Prompted by a request for advice from a graduate who was working for a pet food company, a few years ago I started research on pet food. Fortunately, I obtained the cooperation of many people, won a large competitive research grant from the Japnanese Ministry of Agricuture, Forestry and Fisheries, and successfully developed peptide pet food ingredients. For the details (in Japanese), please see the website (http://foodpeptide.com) of the university-originated venture company Food Peptide Co., Ltd. founded in conjunction with this work.

Humans will sometimes eat foods that "although taste bad appear to be healthful", but pet food that tastes bad will be a commercial failure because pets will not eat it. And even if one replaces a product with a new functional pet food, pet owners will not continue buying it unless some change is manifested. It seems to me that in some ways pet food is, more than human food, chosen according to matter-of-course criteria. I think that "delicious and healthful" is the basis for both functional human foods and functional pet foods. Perhaps it will be possible to apply our experiences in pet food development to human food.

Conclusion : The importance of collaboration

This has been a somewhat broad-ranging paper, but I hope it will serve to spur thought on "food and health". Due in part to my own lack of ability, I was unable to delve into the real "possibilities and limitations" of functional foods, but I think I was able to provide ideas for discussion.

Because this is a symposium on the collaboration of agriculture and medicine, I would like to close with an appeal for the importance of collaboration. Without agriculture-medicine collaboration, there can be no fundamental approach to the issue of food and health. Functional food research and development also require industry-academia-government collaboration. I am employed by a university, and have benefited greatly from collaboration with graduates. I intend to continue carefully fostering those relationships. And needless to say, interaction with the public (consumers) through public lectures and other events is to us a valuable opportunity for collaboration.

Chapter 50

Current State of the Kitasato University Agromedicine Concept

Katsu Minami

Introduction

One major problem in the world now is the "disjunction disease". We see it in the bonds between people, between parent and child, between teachers and students, between soil/nature and humans, between one fact and another, between culture/history and the present, and in countless other examples.

Looking through the examples shows there are four types. The "disjunction of knowledge from knowledge", examples of which are immersion in specialization, the maze of specialized terminology, and the use of non‐alive language. The "disjunction of knowledge and action", examples of which are the disjunction between those who construct theories and those who are responsible for practice, and the disjunction between the virtual and the real. The "disjunction between knowledge and feeling", that is, thoroughgoing objectivism, and the extreme disjunction between knowledge and reality. The "disjunction between past knowledge and present knowledge," that is, the disjunction of the time axis where we learn from history and the passing on of culture, as indicated by terms such as "finding eternal truths in popular trends" and "learning from the past."

At the start of *On the Way of Medicine* (1878) author Shibasaburo Kitasato (1853~931) states his conviction about the healing art. It includes this passage: "The foundation of the way of medicine is to lead the people, have them understand how to care for and protect their health, and with this, to inform them of the value of their physical selves, and hence provide them with the way to prevent illness." To paraphrase this, he is saying, "The basis of medicine is to explain to the people how to maintain their health, tell them about the importance of their bodies, and prevent illness." Another interpretation of this is that it is necessary to eat food that has been produced in a sound environment and produced in safe manufacturing processes, maintain one's health, and not fall victim to illness. Truly, that is the title of this symposium, "Food Safety and Preventive Medicine".

Kitasato continued with scathing criticism of the physicians of his day, which can be paraphrased in this way: "Preventing illness cannot be achieved without a full understanding of medicine; that is, what causes illness and how to cure it. To provide true medical care, one must fully research the healing art. Those who aspire to medical science must exhaustively research both theory and technique, without more weight on one or the other." Another possible interpretation is that before seeing a doctor, people should produce safe agricultural products that prevent illness, and conserve the underpinning environment.

To briefly sum up Shibasaburo Kitasato's *On the Way of Medicine*, he sets forth the conviction that the foundation of medicine is prevention, and says that the achievements of academic inquiry should be used widely to benefit the people. Here we see the idea of practical science that links academic inquiry with practice. There is no disjunction between knowledge and knowledge, or knowledge and practice.

On a trip to Nagasaki for a cholera study, Kitasato used his free time to accurately observe the state of roads, wells, drainage, and other elements of the environment in back streets where people had fallen ill. In relation to liver distomiasis, which is caused by a parasite, he describes the route by which the liver fluke (a flatworm 20–30 mm long, belonging to the order Echinostomida (digeneans)) infects the liver. This achievement was due to his sharp-eyed observations of the environment. He urges attention be directed at sheep that eat the snails that harbor flukes. This is indeed the practical science of Shibasaburo Kitasato, who linked academic inquiry with the real world.

Incidentally, *On the Way of Medicine* concludes with a Chinese-style poem with seven-character lines. The meaning can perhaps be summed up in this way: "If a man endures and braves hardship, there is no reason to expect that he cannot successfully surmount the challenges of public health".

Shibasaburo Kitasato's practical science naturally did not include the disjunction disease. In fact, we must learn from Kitasato's foresight. Agriculture, the environment, and medicine should never have been separated.

With respect to agriculture, the environment, and medicine (that is, agromedicine in an environmental context), who will represent us in examining the desirable state of inter-regional connections within single countries, international connections in the world, interdisciplinary connections among specialties, and intergenerational connections between now and the future? The answer is that those on the forefront are the intellectuals.

But in the near-modern and modern eras, intellectuals are on the decline. In their place there is a growing number of experts who are proficient in certain fields, and that trend is further accelerating amid the advance of our highly information-oriented society. There are few people who give knowledge an integrated interpretation, while those who partially interpret knowledge and use it shrewdly have begun to swagger. What is more, many experts appear to be immersed in their specialties in order to avoid the responsibilities of their fields. The problem of agromedicine in an environmental context is that it is an extremely difficult field for being an intellectual. How should we deal with this situation?

Chapter 11 of the Chinese classic *Dao De Jing* by the "Tao" philosopher Laozi contains this passage.

"We put thirty spokes together and call it a wheel;
But it is on the space where there is nothing that the usefulness of the wheel depends.

We turn clay to make a vessel;

But it is on the space where there is nothing that the usefulness of the vessel depends.

We pierce doors and windows to make a house;

And it is on these spaces where there is nothing that the usefulness of the house depends.

Therefore just as we take advantage of what is, we should recognize the usefulness of what is not. (chap. 11, tr. Waley)"

(Arthur Waley, The Way and Its Power: A Study of the Tao Te Ching and its Place in Chinese Thought, London, Allen & Unwin, 1934, New York, Grove, 1958.)

Here is thought which shows a fundamental principle for unifying diversity. If we interpret this text from the stance of agromedicine, it would mean that to bring about the functions of an "agriculture-environment-medicine" or "food-soil-health" collaboration, it is necessary to use information, education, research, dissemination, and other things each as a useful material in building new rooms.

To aid the search for what agromedicine should be in the future, I shall build on the thinking presented above in describing the state of progress achieved at Kitasato University so far in education, information, research, dissemination, and other agromedicine-related areas.

Publication of "information : agriculture, environment, and medicine"

To promote broad awareness of agromedicine-related information among parties concerned, the university has been sending out "Information: Agriculture, Environment, and Medicine" as a newsletter from the office of the president of Kitasato University. It has been published on the first of every month since the first issue in May 2005, with the 46th and most recent issue in January 2009. Newsletter sections include "Opening Remarks", "University Happenings", "Domestic Information", "International Information", "General Remarks, Resources, and Topics", "Laboratory Visits", "Useful Literature", "Books and Resources", "Lecture Presentations", "People with Agromedicine at Heart", "A Stroll Through Words", "Agromedicine", "Geomedicine", and "Miscellaneous".

Windows, and doors spoken of by Laozi indicate distinctiveness or individuality, while the wheel, vessel, and room indicate the unification of diversity. For example, among the sections of "Information: Agriculture, Environment, and Medicine", the "International Information", "Domestic Information", "Books and Resources", "People with Agromedicine at Heart", and "Faculty Visits" sections correspond to the clay and window. The wheel, vessel, and room have yet to be made. A room called "agromedicine" will of course not suddenly appear or be built.

The agromedicine "room" will only be built after a long time with the interest, cooperation, assistance, effort, and other involvement by many people. If the room is built, though it be a humble one, we can put curtains in the windows, hang paintings, and bring in desks and chairs. In time, we will even get a big sofa for visitors.

Holding the Kitasato university agromedicine symposiums

In addition to "Information: Agriculture, Environment, and Medicine," there is another effort to unify the diversity of agromedicine, and that is the Kitasato University Agromedicine Symposiums. The first was held in March 2006, and we have hosted them at a pace of one every six months. This October 2008 symposium is the sixth. These symposiums perhaps correspond to the door of the agromedicine room.

Holding medicinal plant seminars

With the intent of creating a new urban agriculture, Kitasato University and Sagamihara City have signed an "Agreement on Encouraging a New Urban Agriculture", and are running a collaborative project that involves medicinal plants. As an integral part of this initiative, the partners are working on promoting public awareness of medicinal plants and their use. Additionally, in an effort to heighten interest in new agriculture and to encourage a new agriculture whose focus is on health, the environment, and new urban agriculture, the partners have to date held three "Medicinal Plant Seminars" as an integral part of the liaison between agriculture and medicine.

Education : Holding lectures on agromedical theory reports

Starting with the students who matriculated in April 2007, lectures on agromedicine account for part of the "Theory of Medical Science" lectures for first-year medical school students, and part of three courses for first-year veterinary medicine students: "Introduction to Veterinary Medicine I," "Outline of Animal Science I," and "Outline of Environmental Bioscience I."

In April 2008, we added the course "Agromedical Theory" (one credit) to the College of Liberal Arts and Sciences' "Liberal Arts Seminar B". Professors including those from the School of Medicine, School of Veterinary Medicine, School of Pharmacy, and the Kitasato Institute for Life Sciences take turns delivering lectures.

Under the "High-Quality University Education Promotion Program", we have applied for a project called "Aiming to Build Career Paths Leading to High-Level Professionals through Mastering Composite Knowledge and Technologies".

Content of "liberal arts seminar B : agromedical theory" in the college of liberal arts and sciences

1) Education objective

For purposes including illness prevention, health improvement, safe food, conservation agriculture, and healing agriculture, in other words, for the purpose of making people living in the 21st century happy in both body and mind, it would be impossible to overemphasize the need for agromedical science and education. It is vital that Kitasato University students, who are explorers of the life sciences, are aware of the importance of agromedicine.

In this seminar, students learn basic ways of thinking such as the historical similarity between agriculture and medicine, the agriculture- and medicine-related problems in modern society, agricultural science as seen from medical science, and medical science as seen from agricultural science. The objective is to have students understand real-world phenomena relating to agriculture and medicine in an environmental context, and to learn the necessity of agromedical science.

2) Education content:

Students acquire basic knowledge from courses such as Introduction to Agromedicine, Agromedicine as Seen from the Medical and Agricultural Sciences, Alternative Medicine, and Alternative Agriculture. They also acquire specific global-scale knowledge on avian flu, heavy metal elements, and global warming and agriculture/environment/health. Courses explain the state of agromedical initiatives around the world. There is discussion and examination of the future of agromedicine.

3) Education method:

Lectures and seminars by instructors from various schools and practicing professionals. Lectures use means such as PowerPoint presentations. Students write reports on and discuss lecture series. In final sessions, participants discuss and examine the preferable future of agromedicine.

4) Goals:

Students can understand that in the 21st century the science of agromedicine is essential. They can understand the historical backgrounds of agriculture and medicine. They can see world agromedical trends and see the actual state of agromedical practice.

5) Grading method and criteria:

Reports, tests, attendance, active participation in discussions, and other factors are taken into consideration, and an overall judgment is made.

6) Message to students:

Efforts to solve the problem of how we can link the objectives of medical science in the 21st century with the challenges of agricultural science not only respond to the demands of society but also are vital to Kitasato University, which aspires to be an explorer of the life sciences, and to those who study at Kitasato University.

7) Content of lectures:

1. Introduction to Agromedicine: Katsuyuki Minami

This lecture explains the need for agromedical science while pointing out the historical similarity between agriculture and medicine, and the problems of agriculture and medicine in modern society, and conducts an exchange of views about this.

2. Agromedicine as Seen from Medical Science: Yoshiharu Aizawa

To lead a healthy life in a society that aims to become sustainable, it is crucial to implement nutritional education and nutritional science. The importance of agromedicine is explained along with the lecturer's personal experiences.

3. Agromedicine as Seen from Agricultural Science: Katsuyuki Minami

Food that is capable of maintaining human health can be produced only from soil that has a good balance of nutrients. This lecture introduces the idea that medicine and food spring from the same source, has students understand the need for agromedical science and education, and has them discuss the importance of food.

4. Agromedicine as Seen from Oriental Medicine and Alternative Medicine: Haruki Yamada

This lecture explains Oriental medicine, the official medicine of Japan, while comparing it with Western medicine and recent alternative medicine, and expounds on the current state of and need for agromedicine.

5. Alternative Agriculture: Katsuyuki Minami

Conventional agriculture uses large amounts of pesticides and chemical fertilizers. This has adversely affected crops and the environment. This lecture introduces students to various substitute agricultural practices, explains the need for alternative agriculture, and has students see it in action.

6. Conservation Livestock Farming—From Production to the Hospital Ward: Tomiharu Manda

This lecture introduces students to conservation livestock farming, and describes a system which safe animal products produced in well-conserved environments are used advantageously in hospital wards. The initiatives at Yakumo Farm are also described.

7. Avian Flu—Infection and Control Measures: Shinji Takai

This lecture explains that infection of animals and humans by bird flu is intimately connected to the natural environment and to agricultural production sites, describes how infection happens in the real world, and considers control measures.

8. Avian Flu—Vaccine Measures: Tetsuo Nakayama

This lecture explains ways to prevent infection of people with bird flu, how vaccines are manufactured and used, and other related information.

9. Heavy Metals—A Biogeoscience Perspective and Dealing with the Risk of Soil and Agricultural Product Contamination: Katsuyuki Minami

This lecture explains from a biogeoscience perspective the amounts of heavy metals extracted from the earth's crust and their behavior in the environment, and provides other information such as the processes by which heavy metals are absorbed by the soil and crops, and control measures.

10. Heavy Metals—The Perspective of Clinical Ecology: Ko Sakabe

Health damage caused by heavy metals is a very serious problem in the field of clinical medicine. This lecture explains the history of heavy metals and people, onset mechanisms, and more.

11. Global Warming and Agriculture, the Environment, and Health: Katsuyuki Minami

It is said that global warming is advancing much faster than experts had anticipated. What will happen to agriculture, the environment, and human health because of global warming? The most recent data is used to give the students an understanding of global warming's seriousness, and discusses how to cope.

12. Initiatives and Future of Agromedicine: Katsuyuki Minami

This lecture introduces students to agromedicine initiatives in North America, the Nordic countries, Japan, and other places around the world. This lecture explains concepts such as agromedicine and medical geology. This lecture encourages an exchange of views on matters such as how agromedical science and education should be developed.

Kitasato university symbiosis research :
Research laboratories and research agenda

In the quest for research materials and people capable of linking agriculture, the environment, and medicine, we explored a variety of research departments, starting in April 2005 with Hygienics and Public Health in the School of Medicine and ending in July 2007 with Veterinary Infectious Diseases in the School of Veterinary Medicine. Issues 1 through 28 of "Information: Agriculture, Environment, and Medicine" have information on the 26 courses and research departments of Kitasato University that we visited, including descriptions of the research.

Contributors

Yoshiharu Aizawa
Dean
School of Medicine, Kitasato University

Keizo Arihara
Professor
School of Veterinary Medicine, Kitasato University

Kumiko Asahi
Lecturer
Kitasato Junior College of Health and Hygienic Sciences, Kitasato University

Hideo Harasawa
Director
Social and Environmental System Division
National Institute for Environmental Studies

Yousay Hayashi
Professor
Graduate School of Life and Environmental Sciences, The University of Tsukuba

Hisao Kamiya
Professor Emeritus, Kitasato University

Yutaka Kanai
Nature Conservation Office
Wild Bird Society of Japan

Takafumi Kasumi
Professor
College of Bioresource Sciences, Nihon University

Fujio Kayama
Professor
Faculty of Environmental Medicine,
Center for Community Medicine, Jichi Medical University

Toshiaki Kita
Associate Professor
Kashiwanoha Kampo Clinic
Center for Environment, Health, and Field Sciences, Chiba University

Toyoki Kozai
Professor and Former President
Chiba University

Kikuo Kumazawa
Professor Emeritus, The University of Tokyo

Kazutake Kyuma
Professor Emeritus, Kyoto University
Professor Emeritus, The University of Shiga Prefecture

Tomiharu Manda
Professor
School of Veterinary Medicine, Kitasato University

Anne McDonald
Director
UNU-IAS Operating Unit Ishikawa/Kanazawa, United Nations University

Takeshi Mikami
Chairperson
Food Safety Commission
Cabinet Office Government of Japan

Katsu Minami
Vice-president and Professor
Kitasato University

Takao Mukai
Professor
School of Veterinary Medicine, Kitasato University

Yasuhito Nakagawa
Professor
School of Pharmacy, Kitasato University

Masayuki Nakamura
Professor
School of Veterinary Medicine, Kitasato University

Tetsuo Nakayama
Professor and Control of Viral Infections Deputy Director
Kitasato Institute for Life Sciences, Kitasato University

Hisayoshi Ohta
Professor
School of Allied Health Sciences, Kitasato University

Nobuhiko Okabe
Director
Infectious Disease Surveillance Center
National Institute of Infectious Diseases

Masashi Omura
Lecturer
Kitasato Junior College of Health and Hygienic Sciences, Kitasato University

Shinichi Ono
Head
Soil Environment Division
National Institute for Agro-Environmental Sciences

Hitoshi Oshitani
Professor
Graduate School of Medicine, Tohoku University

Kou Sakabe
Professor
School of Pharmacy, Kitasato University

Masahiro Segawa
Director
Biotechnology Safety Office
Agriculture, Forestry and Fisheries Research Council Secretariat, MAFF, Japan

Tadayoshi Shiba
President and Professor
Kitasato University

Isoya Shinji
Professor and Former President
Tokyo University of Agriculture

Nobutaka Shinohara
Professor
School of Medicine, Kitasato University

Masaki Taga
Lecturer
Kitasato Junior College of Health and Hygienic Sciences, Kitasato University

Shinji Takai
Professor
School of Veterinary Medicine, Kitasato University

Kazuyuki Yagi
Senior Researcher
Department of Global Resource Carbon and Nutrient Cycles Division
National Institute for Agro-Environmental Sciences

Haruki Yamada
Director and Professor
Kitasato Institute for Life Sciences, Kitasato University

Norio Yamaguchi
Professor
Alternative Basic Medicine Course
Kanazawa Medical University Graduate School

Shigeo Yamaguchi
Research Manager
National Institute of Animal Health
National Agriculture and Food Research Organization

Hiroshi Yamauchi
Professor
School of Allied Health Sciences, Kitasato University

Kumiko Yoneda
Senior Research Scientist
Japan Wildlife Research Center

Tadakatsu Yoneyama
Professor
Graduate School of Agricultural and Life Sciences
The University of Tokyo

Yasuhiro Yoshikawa
Professor
Graduate School of Agriculture and Life Sciences, The University of Tokyo

JCOPY <(社)出版者著作権管理機構 委託出版物>		
2009	2009年7月10日　第1版発行	
北里大学農医連携学術叢書第7号		
－農－環境－医療－ （英文）	著作代表者	陽　捷行（みなみ　かつゆき）
検印省略		
ⓒ著作権所有	発　行　者	株式会社　養賢堂 代　表　者　及川　清
定価5250円 （本体5000円 税　5％）	印　刷　者	株式会社　丸井工文社 責　任　者　今井晋太郎
発　行　所	〒113-0033 東京都文京区本郷5丁目30番15号 株式会社 養賢堂　TEL 東京(03) 3814-0911　振替00120 FAX 東京(03) 3812-2615　7-25700 URL http://www.yokendo.com/	
	ISBN978-4-8425-0454-4　C3061	

PRINTED IN JAPAN　　　製本所　株式会社丸井工文社
本書の無断複写は著作権法上での例外を除き禁じられています。
複写される場合は、そのつど事前に、（社）出版者著作権管理機構
（電話 03-3513-6969, FAX 03-3513-6979, e-mail:nfo@jcopy.or.jp)
の許諾を得てください。